World Vegetation Types

WITHDRAWN

World Vegetation Types

EDITED BY
S. R. EYRE

WITHDRAWN

COLUMBIA UNIVERSITY PRESS
NEW YORK
1971

QK
101
E88
1971

Selection and editorial matter © S. R. Eyre 1971

All rights reserved. No part of this publication
may be reproduced or transmitted, in any
form or by any means, without permission.

Published in Great Britain 1971 by
THE MACMILLAN PRESS LTD

ISBN: 0–231–03503–9
Library of Congress Catalog Card Number: 78–147779

Printed in Great Britain

581.526
Eye

WITHDRAWN

Contents

6   *Contents*

List of Plates

Acknowledgements

The Tropical Rain Forest, by P. W. Richards, copyright Cambridge University Press 1952

'La Forêt Coloniale', by A. Aubréville, *Académie des Sciences Coloniales: Annales*, IX, copyright Éditions Maritimes et d'Outre-Mer 1938

'The Natural Vegetation of Trinidad', by J. S. Beard, *Oxford Forestry Memoirs*, No. 20, copyright Clarendon Press 1945

'An Example of Sudan Zone Vegetation in Nigeria', by R. W. J. Keay, *Journal of Ecology*, XXXVII, copyright British Ecological Society 1949

'The Montane Vegetation of New Guinea', by R. G. Robbins © *Tuatara*, VIII 3 (1961)

'The Vegetation of the Imatong Mountains, Sudan', by J. K. Jackson, *Journal of Ecology*, XLIV © British Ecological Society 1956

The Vegetation of New Zealand, 2nd ed., by L. Cockayne (being vol. XIV of *Die Vegetation der Erde*, by A. Engler and O. Drude) (Verlag von Wilhelm Engelmann, Leipzig, 1928)

Plant Ecology, 2nd ed., by J. E. Weaver and F. E. Clements, copyright 1938 by McGraw-Hill, Inc. Used with permission of McGraw-Hill Book Company

'The Broad-Sclerophyll Vegetation of California', by W. S. Cooper, copyright The Carnegie Institute of Washington, no. 319 (1922)

Plant Forms and Their Evolution in South Africa, by J. W. Bews (Longmans, Green & Co., 1925)

'Floristics and Ecology of the Mallee', by J. G. Wood, copyright *Transactions of the Royal Society of South Australia*, LIII (1929)

Guide du Naturaliste dans le Midi de la France, vol. II, by H. Harant and D. Jarry © Delachaux et Niestlé 1963

Agricultural Origins and Dispersals: The Domestication of Animals and Foodstuffs, by C. Sauer © The Massachusetts Institute of Technology Press 1969

Grasslands of the Great Plains, by J. E. Weaver and F. W. Albertson, copyright Johnson Publishing Company, Lincoln, Nebraska 1956

'Vegetation of the Plains of European Russia', by B. Keller, *Journal of Ecology* xv, copyright British Ecological Society 1927

'The Vegetation of Alberta: IV. The Poplar Association', by E. H. Moss, *Journal of Ecology*, xx, copyright British Ecological Society 1932

'Climatic Change or Cultural Interference? New Zealand in Moa-hunter Times', by K. B. Cumberland, *Land and Livelihood: Geographical Essays in Honour of George Jobberns* © New Zealand Geographer 1962

The Grasses and Grasslands of South Africa, by J. W. Bews (P. Davis & Sons Ltd, Pietermaritzburg 1918

The Changing Mile, by J. R. Hastings and R. M. Turner © The University of Arizona Press 1965

Nature of Sinkiang and Formation of the Deserts of Central Asia, by E. M. Murzayev © Joint Publications Research Service No. 40, 299 (1967)

Introduction to Plant Geography, by N. Polunin © Longmans, Green & Co. 1960

Introduction

GENERALLY speaking, the study of the distribution of plant life over the earth has attracted the attention of two kinds of investigator, though this fact has been obscured by the terms that have been used to refer to their fields of study. On the one hand there are those who have been primarily concerned with the individual species and with *taxonomically* related groups of species; on the other there are those who have been preoccupied mainly with the groupings of taxonomically diverse species into *plant communities*. The former have been taxonomic botanists, despite the fact that they have often referred to their field of study as 'plant geography'; their interests have been focused on the problems of divergence, speciation and dispersal – they have thought in terms of distance, migration barriers and discontinuous distributions. The latter are now usually referred to as 'ecologists', though they too have often been happy to accept the title of 'plant geographer' or 'phytogeographer'. They are concerned with the relationships between plants and the physical environment and with the complex interactions between the plant and the other organisms within the same ecosystem.

These two fields of study, though distinct, are obviously not mutually exclusive. It is true that taxonomists have, on occasion, been heard to exclaim that their interest is in plants and not in vegetation, but on the whole this declaration of respectability has been good-humoured, with no more than the merest hint of malice! Indeed many taxonomists, from Darwin and Hooker onwards, have shown an obvious awareness of the effect of the community itself on its component individual plants. They are increasingly aware of the necessity for an assessment of the nature of the physical and biotic environments in which species have evolved, just as ecologists know of the benefits to be gained from an understanding of the ways in which emerging species have expanded their territories and intruded into established communities.[1]

[1] As instanced by the recent joint meeting of the Linnean Society and the British Ecological Society (October 1968) to discuss speciation in the tropical environment. See R. H. Lowe-McConnell (ed.), *Speciation in Tropical Environments* (Linnean Society of London, May 1969).

It would be untrue to say that geographers are in no way concerned with the 'plant geography' of the taxonomist. Amongst other things, they are interested in the origin and spread of cultivated plants; indeed the author of one of the passages in this anthology has been one of the most stimulating and original thinkers in this field.[1] Nevertheless, despite the term 'plant geography', it is the botanist, inevitably, who has been pre-eminent in this aspect of plant study. Geographers, because of their primary interest in the interactions and interconnections within the environment as a whole, have always been more directly concerned with plant communities and with plant ecology. This fact has remained obscure, however, mainly because of the protracted usage of the time-honoured expression 'natural vegetation'.[2]

Since the present volume is one of a geographical series, it thus seems appropriate that it should be concerned with the nature and distribution of plant communities rather than with particular plants. The vegetation cover of the earth is almost infinitely variable, however, so that any small collection of writings about it must be highly selective. The editor's problems must be quite apparent; nevertheless an apology must be made at the outset to all those readers who seek here for passages about the vegetation of particular areas only to be disappointed. It has only been feasible to present a representative selection of writings which deal with examples of those *main types of wild vegetation* which covered most of the land areas of the earth until relatively recent times.

There is no concise scientific definition of the expression 'main vegetation type'. Many of the entities which have been portrayed on traditional atlas maps of 'natural vegetation' are certainly 'climaxes'

---

[1] Carl O. Sauer. See *Agricultural Origins and Dispersals*, Bowman Memorial Lectures (American Geographical Society, 1952).

[2] The exact meaning of this term was never completely clear, and with the development of the notion of 'climax' over the past half-century it has become positively ambiguous: it is not synonymous with the term 'climatic climax' since a number of vegetation categories shown on maps entitled 'Natural Vegetation' are not entirely climatic climax formations; on the other hand no other satisfactory definition for it has been suggested. This term of convenience appears to have become fossilised in geographical teaching because the vegetation belts of Humboldt, de Candolle and Schimper were presented as 'natural vegetation', each type enshrined in its own climatic region. The level of ecological thinking in geography has suffered in consequence. See S. R. Eyre, 'Determinism and the ecological approach to geography', *Geography*, XLIX (1964) 369–76.

or 'plant formation-types' in the classical terminology of Clements[1] and Tansley:[2] They are true climatic climax groupings, each dominated by plants of the same life-form. Other widespread types are of a very different kind, however: in the eighteenth and nineteenth centuries very large areas were covered with vegetation whose ecological status is now in question. Although Schimper[3] appears to have been quite uninhibited in accounting for their existence in fairly simple climatic terms, communities such as tropical savanna and the true prairie of middle latitudes are now recognised as posing problems of ecological status which are most intriguing in their complexity.

Because of these problems, purely descriptive writings about vegetation can never again be entirely satisfying. This is not a collection of writings about the nature of the ecosystem or any other aspect of ecological theory; nevertheless it is felt to be imperative to include only passages which explore developmental aspects of plant communities as well as describing their structure and composition. Furthermore, since certain communities such as the wild grasslands already mentioned have given rise to much controversy, they have been given coverage here which is quite disproportionate to their areal extent. On the other hand certain types of forest have been neglected. This should not be taken to imply that the nature of every single plant community does not repay prolonged and careful study – merely that some cases seem to illustrate problems in development or the effects of human interference more strikingly than others.

No excuse is offered here for an underlying acceptance of the Clementsian philosophy based on the concept of climatic climax. Investigations over the last few decades may well have demonstrated that plant formations and formation-types are by no means such distinct entities or in so stable a state of equilibrium as was postulated half a century ago; the inevitable ultimate dominance of a particular life-form under a given set of environmental conditions is certainly open to doubt. Indeed some of the extracts presented here demonstrate clearly that such sweeping views never did find universal acceptance. The fundamental idea of *succession* remains, however, and it is difficult to conceive of an ecological philosophy which does not

[1] F. E. Clements, 'Nature and structure of the climax', *J. Ecol.*, XXIV (1936) 252–84.

[2] A. G. Tansley, *The British Islands and Their Vegetation* (1939; reprinted 1949) I 228–32.

[3] A. F. W. Schimper, *Plant Geography upon a Physiological Basis*, trans. W. R. Fisher (1903).

embrace it in some form. Similarly it is difficult to see how the great
similarity of life-form found in taxonomically diverse groups in the
same type of environment can possibly be explained without recourse
to the notion of convergent evolution. With these points in mind, it
still seems appropriate to arrange an anthology on world vegetation
on the basis of plant formation-types.

It might be argued that, at the present stage of world 'development'
it is somewhat unrealistic (possibly even rather romantic) to devote a
volume such as this to wild plant communities the like of which are
difficult if not impossible to find. The great forests and grasslands
have been swept away, ploughed up or, at least, very much modified
by man's activities. Is there not a good case for a more up-to-date
collection of writings in which a representative sample of the man-
induced or plagioclimax communities of the earth are discussed?
Indeed, would it not be more in keeping with the spirit of the
twentieth century to devote the work entirely to articles on the ecology
of the tea plantation, the orange grove, the prairie wheat stand or the
ryegrass–clover meadow?

While one must acknowledge the reasonableness of these suggestions, and indeed the complete feasibility of their achievement, the
attitude of mind which would maintain that such collections of
writings are the only ones which are worthy of attention would be a
deplorable one. The effect of human activity on wild vegetation cover
is, of course, a topic of prime importance – in a sense it is all-
important. Indeed, quite a number of the extracts here are very much
concerned with it. But one must not infer from this that the nature of
the original cover and the ways in which it was reduced or transformed
are things which can now be completely ignored and forgotten. To do
so would be to put oneself in the position of a driver at high speed on
a lonely motorway who becomes so mesmerised by the speed with
which the rest of the earth slips past him and by the straightness of
his trajectory that he develops the illusion that constant high speed
in a constant direction is the normal state of existence; the nature of
the starting point is forgotten and an awareness of the importance of
brake and steering wheel is lost. We must not forget that our present
croplands, plantations, heaths and meadows are not in a state of
'normalcy'; in fact they are very new and very 'abnormal' when
viewed in the light of the age of the earth's great ecosystems. Quite
apart from wheat fields and rubber plantations – the products of our
conscious manipulations – there are large areas of wild grassland,

savanna and scrub which have been produced inadvertently by human activity. None of these can be fully understood merely by visual examination, or even by thorough analysis of their regimes and chemical turnover. All the land areas of the earth have biological potentialities and, in order to assess them, experimentation and analysis are certainly important; but to come to the fullest possible understanding of these potentialities it is necessary to have some knowledge of the original climax situation. This must surely remain the first premise for 'production ecology' even though, in some areas, one may reach the ultimate conclusion that nature was perhaps not quite so efficient as she might have been, and that introduced species and combinations of species can increase biomass production or biochemical variety. It seems reasonable to expect that a more critical view of the effects of man and of the contemporary situation will be forthcoming from one who has tried to visualise and reconstruct the untamed starting point (difficult though this may be) than from one who merely dissects the contemporary and becomes prone to regard it as 'normal'. If one may be permitted a parallel from the general sphere of human affairs, present-day society might well assess its ills with more realism if it set store by history to the extent that it espouses psychology and sociology.

From the point of view of animal ecology and conservation also, a knowledge of the nature of the original vegetation cover is of the utmost importance. Many of the more successful species of wild animal have flexible habits and are therefore able to adapt themselves as the environment is changed by human activity; indeed a minority of species have profited enormously from the advent of plantations and field crops. A very large number, however, are so specialised and intimately adapted to specific features of the wild environment that the slightest interference with this may strike a mortal blow at the very roots of their existence. In all those areas where human communities have modified the vegetation in any way, it is impossible to think realistically about the present density and distribution of animal populations without first acquiring a knowledge of the original vegetation cover in which the species concerned evolved and subsisted for countless generations.

SCOPE AND CONTENT

The editor has therefore been influenced by two main considerations in his selection of writings for this anthology. The first aim has been to present a set of studies which are fairly representative of those main types of wild vegetation which, until recently, covered much of the land area of the earth; the second has been to include only that kind of work which takes cognisance of ecological status and vegetation development. Ecological investigation of this kind usually involves close scrutiny not only of natural environmental factors but also of the history of land utilisation by man; indeed an examination of most of the extracts here will confirm that conclusions regarding the original nature of the wild vegetation of an area cannot usually be regarded as valid unless careful consideration has been given both to natural factors and to past human activity. The very nature of this kind of study makes it inherently geographical and, as will be seen from the list of authors, notable contributions have been made by geographers.

Apart from the influences and interactions of human and natural factors, however, other quite distinct problems are encountered, particularly when analysing the composition and distribution of forest communities. Foresters such as Aubréville and Cockayne have postulated that the actual processes of regeneration may well create recognisable patterns which are quite independent of the factors of the macro-environment, even in forest which would normally be regarded as stable climax. This appears to be fundamental; consequently it is felt that such developmental phenomena must receive some attention in an anthology which seeks to take an overall view of world vegetation.

It will be clear to the reader that the extracts included here are the products of two or three generations of ecological thought. Although a few of them were written some half a century ago, their inclusion has seemed imperative. Consequently some of the ecological theory found here may no longer be entirely acceptable, but these earlier writers had the advantage of being able to examine little-altered vegetation communities which have now disappeared. Furthermore they lived in a period when careful and detailed description of plant communities was still a major preoccupation. Even though a plant community may have survived more or less intact into the latter half of the twentieth century, it is not always possible to find a recent

581.526
Ey6

description of it which attempts to be comprehensive. On the other
hand the editor has resisted the temptation to include passages from
the stimulating classical descriptions of vegetation which emanated in
such abundance from the explorer-naturalists of the eighteenth and
nineteenth centuries. The colour and wild grandeur which leaps from
the pages of Humboldt, Berg, Hochstetter, Wallace and so many
more is something which, with regret, one must forgo. Although these
great naturalists show frequent flashes of profound ecological in-
tuition, they were writing before general ecological theory, on a
developmental basis, had been formulated; they provide descriptive
writing at its best, but in general they could not be expected to discuss
the composition and distribution of plant communities at a level
which could today be regarded as useful. Within the restricted space
of this anthology it has been necessary to select writings which, in
the main, attempt the functions of both description and interpretation; a collection of extracts which date from the period between the
third and seventh decades of the twentieth century seems to provide
this blend most satisfactorily.

TROPICAL PLANT COMMUNITIES

The plant communities of inter-tropical regions receive attention
here before those of higher latitudes. This seems realistic from the
evolutionary point of view. The evidence suggests that, generally
speaking, it is plants of tropical origin that have adapted themselves
to the cooler parts of the earth, rather than the reverse. The reasons
for this are apparent when one remembers that, in Mesozoic and
early Tertiary times, much warmer climatic conditions were to be
found in middle latitudes than is the case today, and indeed forest
communities very similar to those now found in the tropics and
moist subtropics extended as far north as the London Basin,[1] if
not further. There may well have been a protracted period when
forests with a structure and combination of life-forms quite similar to
those now found in the tropical rain forest covered the major part of
the land areas of the earth. It seems appropriate therefore that some
account of the structure and life-forms of present-day tropical rain
forest should have first place here, and equally fitting that the extract
should be taken from *The Tropical Rain Forest* by P. W. Richards.
 This important plant formation-type has a number of significant

[1] E. M. Reid and M. E. J. Chandler, *The London Clay Flora* (1933).

aspects quite apart from its structure and physiognomy, and probably the most fundamental of these is its mode of regeneration. Mention has already been made of the general significance of forest regeneration processes; in fact it is in the tropical rain forest that these regeneration phenomena are most clearly instanced. They have far-reaching implications regarding the basic association pattern as well as in the study of 'secondary forest' development, microclimatic fluctuations, faunal contrasts and so on. Clearly they merit special attention. Many views are held regarding the exact nature of regeneration in virgin forest, but probably the most interesting one was put forward by Aubréville in 1938. This may have very wide implications for forest ecology generally, and a translation has been made of part of the work for inclusion here.

Moving from the perennially wet regions into those parts of the tropics where summer rains alternate with winter drought, or where rainfall is generally less regular and reliable, a wider variety of wild plant communities is to be found. These include a number of types of closed forest in which a greater or lesser percentage of deciduous trees are dominant. Extracts from the monograph by J. S. Beard on the vegetation of the island of Trinidad have been selected to present a picture of these types of formation.

Much more widespread in these seasonally dry areas, however (and indeed extending into the perennially wet regions in places), are those plant communities in which grasses and other herbaceous plants form an almost continuous cover over the ground. They are often accompanied by a discontinuous scattering of trees but are sometimes entirely treeless. The very existence and nature of these savannas, occurring as they so often do in close proximity to true forest communities, continues to pose most difficult ecological problems to which there can be no simple answer. Nevertheless, many ecologists and foresters feel that there is overwhelming evidence indicating the influence of human communities on the present vegetation: they see domesticated herds and recurrent fires, particularly the latter, as the main reason for the predominantly grassy nature of the tropical savannas. The application of this generalisation to any particular area, without careful investigation, is, of course, just as hazardous as the application of any other broad generalisation in ecology; meticulous examination of all the available local evidence regarding the history of land tenure and land utilisation is always necessary quite apart from systematic observations of seral trends. Numerous careful investigations of this nature have now been carried out, one good example being the work of R. W. J. Keay in northern Nigeria, an extract from which is included here. Since much more discussion of the effects of fire on vegetation appears in later extracts on the mid-latitude grasslands, it has been thought unnecessary to elaborate this theme further in the savanna context.

It would be misleading, however, to give the impression that ecologists are completely agreed that human activities (along with natural fires perhaps) are the only causes of savanna. There is some evidence to support the view that many ancient peneplains in the tropics may be incapable of carrying thick, closed-canopy forest even though the climate they experience would lead one to expect such a vegetation cover.[1] It may well be that these areas, with their deeply weathered and heavily leached regoliths, can only support an undemanding type of vegetation in which an important element of heliophilous grasses can thrive, and that there is no need to postulate frequent fires as a means of reducing tree cover. Up to the present time, however, investigations along these lines, though most stimulating, remain at a speculative and correlative level; much detailed work is needed concerning the rooting systems, nutrient cycles and hydrological regimes in the plant communities on these ancient surfaces before any firmer conclusions can be reached.

It has seemed appropriate also to include here two passages on the vegetation sequences to be found on tropical mountains. Though of limited extent, these are the only places on the earth where it is possible to pass from tropical heat to frequent frost within a distance of only a few miles. The close juxtaposition of plant communities which results is of absorbing interest; indeed the opportunity to study the zone of interaction between distinct communities in a narrow ecotone is provided here to an unparalleled degree. R. G. Robbins has produced a most lucid and concise account of a sequence to be found in New Guinea, where high mountains rise from a wet tropical lowland; J. K. Jackson has written a comparable account of the vegetation on the Imatong Mountains in southern Sudan where mountain slopes rise from a drier type of tropical environment.

A word of explanation is necessary regarding the absence from this

---

[1] Monica M. Cole, 'Cerrado, Caatinga and Pantanal: the distribution and origin of the savanna vegetation of Brazil', *Geographical Journal*, cxxvi (2) (1960).

anthology of any account of the vegetation of the tropical sides of the perennially dry regions of the earth. Although the plants and plant communities of tropical deserts and semi-deserts possess some unique characteristics which merit separate study, nevertheless they have many characteristics in common with the truly xerophytic communities of higher latitudes. Furthermore, a number of taxonomic groups such as the Cactaceae and Euphorbiaceae have many genera which are common to both. Although several excellent accounts of the vegetation of tropical deserts could well have been selected for inclusion here, it has been decided to concentrate attention on the equivalent xerophytic communities in subtropical and mid-latitude areas. Generally speaking these have been studied more intensively.

PLANT COMMUNITIES OUTSIDE THE TROPICS

European civilisation developed in a forested landscape. With few exceptions the agricultural communities here evolved against a background of trees, from the evergreen oaks of the Mediterranean lands to the deciduous oaks of Sherwood Forest and the beeches on the mountain slopes of central Europe. Furthermore, as Europeans have spread out in middle latitudes during the past five centuries, they have established themselves most quickly and successfully in forested areas: the nations of North America, Australia, New Zealand and South Africa have expanded from forested coastal foci.

Peoples speaking European languages thus have an intimate knowledge of many kinds of mid-latitude forests; consequently there is a vast literature on the subject and any selection of extracts must, of necessity, be rigorous and somewhat arbitrary. It seems very appropriate, however, to select a passage from Cockayne's works on the forests of New Zealand. Not only was this eminent ecologist and forester able to examine the whole range of New Zealand forest types at a time when substantial areas of all of them were still little exploited, he was also dealing with plant communities which had survived from Tertiary times without suffering the drastic dislocations and modifications which have been experienced during the Pleistocene by their counterparts in the northern hemisphere. In Miocene times most moist areas in middle latitudes were covered with mixed evergreen forest – partly broad-leaved angiosperm but partly coniferous

– similar in general form to those which clothed much of New Zealand until the nineteenth century.

From the standpoint of forest evolution, therefore, it seems logical that an account of New Zealand forest should precede any treatment of forest in the northern hemisphere. Here the great variety of types makes selection difficult: both the broad-leaved angiosperm and the needle-leaved coniferous are widespread, and both of these may be either evergreen or deciduous. Furthermore, all these four life-forms can occur in varying proportions in mixed stands. It is only after much hesitation that the very clear and concise account of the coniferous forests of North America, extracted from the work of Weaver and Clements, has been selected. An account of broad-leaved forest could equally well have been chosen; nevertheless these coniferous communities, often very monotonous in their species content, provide a good example of the types of forest which reoccupied the glaciated and periglacial types of surface after the ice retreated not so very long ago.

It may appear peculiar that a collection of writings which is world-wide in scope and yet so limited by space should include no less than four passages which deal with the sclerophyllous vegetation of 'mediterranean' climatic regions. This apparent inconsistency is due to the desirability of presenting at least one major instance of *convergent evolution*. The concept, central to the Clementsian notion of 'life-form', is particularly well illustrated by the sclerophyllous formations. The great physiognomic similarity between the chaparral of California, the Cape sclerophyll of South Africa, the mallee of Australia and the sclerophyllous trees and shrubs of Mediterranean Europe is very well shown by these extracts from Cooper, Bews, Wood, and Harant and Jarry respectively. The first three are typical pieces of ecological writing dating from the early part of this century; all three writers were content to devote a great deal of time and energy to careful physiognomic and floristic description, though, quite clearly, all had come to accept the tenets of developmental ecology and presented their subject matter in the light of this. The fourth passage is of a rather different nature, not merely because it is more modern, but because it deals with an area where it is almost impossible to find a plant community which has not been profoundly affected by direct human interference – usually over a long period of time. Clearly there are many sclerophyllous trees and shrubs which are native to the south of France, but the formulation of a convincing, reasoned argument which places these into communities, and the communities into a hierarchy according to ecological status, has demanded prolonged research by French ecologists. Harant and Jarry present the fruits of that research in a most interesting fashion and the editor is convinced that no apology need be made for selecting and translating a passage from this semi-popular work.

A number of significant facets of the sclerophyllous scrub formation-type are highlighted in these four extracts. One notices, for instance, the importance of different species of evergreen oak (*Quercus*) in both of the northern hemisphere formations. Also, Wood's work on the mallee shows clearly that while this is floristically transitional between the formations which flank it to north and south, it is physiognomically quite distinct from both of them. Specific points such as these emphasise the real existence of this formation-type.

Although the following six extracts may appear to concentrate rather narrowly on one type of community – the mid-latitude prairies and steppes – they are, in fact, intended to illustrate a problem of the greatest significance. Grasslands are certainly to be found in many of the drier areas in middle latitudes, but whether the grass life-form is the one which is the best endowed to become dominant here has been seriously questioned. At one extreme there are those who doubt the very existence of pure climatic climax grassland; they feel that the invasion of trees and shrubs has only been precluded by frequent firing. At the other extreme there are ecologists who find grass communities such as the American mixed prairie to be so closely integrated and well established that they cannot seriously question their status.

A few paragraphs of the forceful prose of Carl Sauer are used to introduce the topic here – paragraphs in which he outlines so clearly the obstacles to his acceptance of climatic climax status for mid-latitude grassland. Juxtaposed is a passage from the work of J. E. Weaver and F. W. Albertson in which all the emphasis is placed upon the close integration of the mixed prairie of the Great Plains. To these latter authors all the evidence shows that the component species of this type of community are so interdependent that it is impossible to consider it as anything but climatic climax. They barely mention an alternative! The following four extracts from Keller, Moss, Cumberland and Bews deal with aspects of grassland ecology in the Russian steppes, the Alberta prairies, the Canterbury Plains and the South

African veld respectively. All these authors emphasise the importance of fire as an ecological agent; they are clearly agreed on its
efficiency in preventing forest invasion of grassland. It is instructive,
however, to study the different emphases to be found in their work:
both Keller and Moss feel bound to leave open the question as to
how far forest would be capable of spreading into grassland were it
not for frequent fires; Bews seems to distinguish quite definitely
between plagioclimax grassland in the moister areas of the veld and
climatic climax in the drier areas, though he makes no firm suggestion
regarding the position of the boundary; Cumberland, on the other
hand (following Zotov *et al.*), goes so far as to postulate a division of
the Canterbury Plains into original ('true') grassland areas and areas
which were cleared of forest by man-made fires. The nature of the
evidence, and the range of inference which has been drawn from the
evidence, are well exemplified in these six extracts. Clearly it is one
of the most absorbing topics in the whole field of vegetation studies,
as well as being the one which best illustrates how effective human
activities may have been in determining the nature and distribution
of what was formerly taken to be the 'natural vegetation' of the
earth.

As one approaches the environmental extremes, in terms of both
cold and water availability, the absence of trees on the landscape
cannot be attributed to anything but natural environmental stresses.
In both the dry semi-deserts and the arctic tundras, woody plants
certainly abound, but they are more or less low-growing and intermixed with greater or lesser proportions of succulents, grasses and
other herbaceous plants. The ecological significance of these mixtures
of life-forms has given rise to much speculation and serious investigation. Two passages on the vegetation of semi-arid lands are included
here: Hastings and Turner have produced an incisive account of the
plant communities of the Sonoran region between 25° and 35°N.
in western North America, while Murzayev has described the distribution of plants in the rather different but equally dry environment
of Sinkiang between 35° and 45°N. in central Asia.

The variety to be found in the vegetation of arctic tundra is just
as great as that in semi-deserts. Grasses, broad-leaved herbaceous
plants, deciduous dwarf shrubs and heathy shrubs have all had some
degree of success during the reoccupation of these northernmost
lands since the ice retreated only a few thousand years ago. The final
extract in this anthology is from the work of Nicholas Polunin, a

leading authority on the arctic tundra; it indicates very clearly some of the variety to be found in tundra vegetation.

No matter how authoritative and carefully chosen, a mere twenty-one extracts cannot provide an adequate picture of the earth's vegetation, nor can they present a comprehensive view of all the major controversies that have arisen regarding the composition, extent and ecological status of all the major types of plant community. It is only hoped that this collection of writings will demonstrate that vegetation study is concerned with evolving, flexible and often elusive entities, the nature of which must be pondered by anyone who wishes to be taken seriously as an authority on land use and on the conservation and manipulation of the earth's biological resources. In the wider sphere, it is certainly hoped that these collected extracts will give pause to those who may remain so uninhibited as to be capable of referring, without qualm, to man's 'exploitation' of these resources; more specifically, it is hoped that students of geography will be helped towards a full realisation that an examination of the distributions to be found on a map of 'natural vegetation' is less than adequate.

1 The Structure of Tropical Rain Forest: Synusiae and Stratification

P. W. RICHARDS

From *The Tropical Rain Forest* (Cambridge University Press, 1952; 1964 impression) pp. 19–34.

THE SYNUSIAE

A COMPLEX plant community is analogous (though admittedly only superficially) to a human society. The members of a human community form social classes, all the members of a given class standing in a similar relationship to the members of other classes and having a similar function in the society as a whole. Each human community thus has a characteristic social structure determined by the nature and the relative importance of the classes which compose it. In a like fashion the species in the more complex plant communities form ecological classes or groups. In the community as a whole the species are of varied stature and varied life-form, but the members of the same ecological group are similar in life-form and in their relation to the environment. These ecological groups, the analogues of the human social classes, will here be called synusiae, a term originally introduced by Gams [1]. A synusia is thus a group of plants of similar life-form, filling the same niche and playing a similar role, in the community of which it forms a part. In the words of Saxton [2], who used the term in a slightly broader sense than Gams, it is an aggregation of species (or individuals) making similar demands on a similar habitat. The species of the same synusia, though often widely different taxonomically, are to a large extent ecologically equivalent.

In a plant community such as the temperate Summer forest the component synusiae are readily distinguished and have a relatively simple spatial arrangement. Commonly the woody plants form two layers or strata, one of tall trees, another of shrubs or smaller trees, while below these there are one or more layers of herbs and undershrubs (the ground or field layers) and a layer of mosses and liverworts. Besides these layers of self-supporting plants, each of which can be regarded as a separate synusia, there are climbers and epiphytes (the latter as a rule consisting entirely of non-vascular plants) and the synusia of saprophytes and parasites, which include a few flowering plants, as well as fungi and bacteria.

In the Tropical Rain forest the synusiae are more numerous and their spatial arrangement is far less obvious. Though tropical forests, like temperate, have a definite structure, they appear at first as a bewildering chaos of vegetation; in the often quoted phrase of Junghuhn, nature seems to show here a *horror vacui* and to be anxious to fill every available space with stems and leaves. A closer study shows, nevertheless, that in the Rain forest, as in other complex plant communities, the plants form a limited number of synusiae with a discernible, though complicated, arrangement in space. This arrangement is repeated as a pattern with only a small amount of variation throughout the Tropical Rain forest formation-type. The structure of primary Rain forest in America is essentially the same as in Africa or Asia. In all three continents the forest is composed of similar synusiae and, as will be seen later in this chapter, there is great similarity in their spatial arrangement. Species of corresponding synusiae in different geographical regions, as well as being alike in life-form, are to a considerable extent alike in physiognomy. The fundamental pattern of structure is thus the same through the whole extent of the rain forest.

Our knowledge of the rain-forest flora is still so imperfect that a complete biological spectrum for even a single limited area of rain forest cannot be constructed (whether the epharmonic life-form system of Raunkiaer is adopted or a purely physiognomic system such as that proposed by Du Rietz). Since the synusia is a group of plants of similar life-form, it follows that a final and complete classification of the rain-forest synusiae is not at present possible. Some kind of classification is, however, a practical necessity, and for the time being we must be content with an approximate and over-simplified scheme. That which will be adopted here is as follows; it is a modification of the scheme used by the author in previous works and has the merit of being convenient and easily applied.

Synusiae of the Tropical Rain Forest

A. Autotrophic plants (with chlorophyll)

1. Mechanically independent plants
    (*a*) Trees and 'shrubs' } arranged in a number of strata
    (*b*) Herbs   } (layers)

2. Mechanically dependent plants

    (*a*) Climbers

    (*b*) Stranglers

    (*c*) Epiphytes (including semi-parasitic epiphytes)

B. Heterotrophic plants (without chlorophyll)

    1. Saprophytes

    2. Parasites

The basis of this scheme, as will be seen, is the means by which the plant satisfies its carbohydrate requirements. Each synusia represents a different method of succeeding in the struggle for food.

The autotrophic plants, which manufacture their own carbohydrates, are directly dependent on light for their existence and are divided into two groups according to their method of reaching it. The mechanically independent, or self-supporting, plants reach the light without assistance from other plants; the mechanically dependent plants (equivalent to the guilds or *Genossenschaften* of Schimper) cannot do so. The independent plants can be subdivided into several strata or layers, the number and structure of which are discussed below.

The dependent plants, on the other hand, are less clearly stratified; such tendency to stratification as they show is determined mainly by that of the independent plants which support them. One section of them, the synusiae of climbers, consists of plants which are rooted in the ground, but need support for their weak stems. Another, the epiphytes, are more or less short-stemmed plants which are intolerant of conditions at ground level and need to grow raised up and rooted on the stems and branches of other plants. One group of these, the semi-parasitic epiphytes (represented by the single family Loranthaceae), obtain water and mineral matter (though probably little or no organic food) from the trees on which they grow; the majority, however, depend on the supporting plant for mechanical reasons alone. The third section of dependent plants, here termed stranglers, are a transitional group. They begin life as epiphytes and later send roots to the ground; some of them reach a considerable size and may ultimately kill and supplant their 'host', as do, for example, the strangling figs (*Ficus* spp.). Stranglers thus begin life as dependent plants, but eventually become independent; they cannot be sharply separated from hemi-epiphytes, such as many Araceae, which send down roots to the ground, but never become self-supporting.

Lastly, there are the saprophytes and parasites, which together form the class of heterotrophic plants. They are not necessarily unresponsive to light, but as they obtain their organic food directly or indirectly from other plants, they are not dependent on it for their nutrition; they have, as it were, retired altogether from the struggle for light.

As far as is known there is no climax community to which the term Tropical Rain forest can properly be applied in which all of these ecological groups are not present. There is also no region of the Rain forest which possesses local synusiae not found elsewhere. The variations in the plan or pattern of structure which are met with consist chiefly of differences in the relative strength with which the various synusiae are represented. Thus in some rain-forest communities epiphytes are represented by more species and individuals, and show a wider range of form, than in others. Similarly, some Rain forests are richer than others in climbers. The spatial arrangement of the synusiae, which must now be considered, also varies to a greater or lesser extent.

STRATIFICATION

The plan of structure common to all climax Tropical Rain forest is most clearly manifested in the central feature of its architecture, the stratification of the trees, shrubs and ground herbs. It has long been a commonplace of the textbooks that in tropical forests the trees form several superposed strata (the terms layer, story, canopy and tier are also used), while in temperate forests there are never more than two tree strata and sometimes only one, but the statement has been very variously interpreted and until recently its precise meaning has been far from clear. Sometimes it is stated categorically that there are three tree strata in the rain forest (according to a few authorities, more than three). Brown [3] for example, describes the stratification of the Philippine Dipterocarp forest in these words: 'The trees are arranged in three rather definite stories. The first, or dominant, story forms a complete canopy; under this there is another story of large trees, which also form a complete canopy. Still lower there is a story of small scattered trees.' Some writers give the impression that the strata of the Rain forest are as well defined and as easy to recognise as in an English coppice-with-standards, but Brown continues: 'The presence of the three stories of different trees is not evident on casual observation, for the composition of all the stories is very complex and few of the trees present any striking peculiarities while smaller trees of a higher story always occur in a lower story and between the different stories.'

There are also authors who state, or imply, that any grouping of the trees according to their height is arbitrary and that 'strata' have no objective reality. This is the point of view of Mildbraed [4], who, with special reference to the forest of the southern Cameroons, says explicitly: 'It is often stated that the Rain forest is built up of several tiers or stories. These terms may easily give a wrong idea to anyone who does not know the facts at first hand. What is meant is merely that the woody plants can be grouped into 3, 4, 5 or perhaps more height classes, according to taste. The space can indeed be thought of as consisting of height intervals of 5, 10 or 20 metres, and species are found which normally reach these height intervals when full grown. As, however, trees of all intermediate heights are present and the mixture of species is so great, these hypothetical height intervals never really appear as stories. It is truer to say that the whole space is more or less densely filled with greenery' (transl.). A similar opinion appears to be held by Chevalier [5] and others.

In this chapter it will be shown that in most normal primary rain-forest communities tree strata actually exist and that they are usually three in number. The tree strata, though always present, are ill defined and are seldom easy to recognise by casual observation. In addition to the tree strata, there is a layer of 'shrubs' and giant herbs (such as tall Scitamineae) and a layer of low herbs and undershrubs.

By a stratum or story is meant a layer of trees whose crowns vary in height about a mean. In a several-layered forest community each stratum will have a distinctive floristic composition, but since the forest is continually growing and regenerating, a proportion, perhaps the majority of the individual trees in the lower stories, will belong to species which will reach a higher stratum when mature. The crowns of young trees of a higher stratum and exceptionally tall ones of a lower will also be found between one stratum and the next; if such individuals are numerous, as is often the case, there will be little vertical discontinuity between neighbouring strata and the stratification will be much obscured. A tree stratum may form a continuous canopy or it may be discontinuous; that is to say, the crowns may be mostly in lateral contact with one another or they may be widely separated. The term canopy is sometimes used as a synonym for stratum; a canopy means a more or less continuous layer of tree crowns of approximately even height. The closed surface or roof of the forest is sometimes loosely referred to as the 'canopy'; in the Rain forest, as we shall see, this surface may be formed by the crowns of the highest tree stratum alone or (more frequently) by the highest and the second stories together.

The divergences of opinion as to the existence, nature and number of the rain-forest strata are mainly due to the difficulty of obtaining a clear view of the forest in profile. On the floor of the forest the observer is, as Mildbraed says, a prisoner. Visibility is very limited both vertically and horizontally. Above, the confusion of leaves, twigs and trunks only occasionally allows a glimpse of the crowns of the taller trees. Here and there, it is true, a gap in the canopy made by the death of a large tree will reveal rather more, or a river or a clearing will make it possible to see a whole cross-section of the forest from top to bottom. Such cross-sections, however, may be extremely misleading. When increased light is able to reach the lower levels of the forest, a vigorous outburst of growth takes place in the undergrowth, climbing plants grow down from the canopy and up from the ground, till the exposed surface becomes covered with a dense curtain of foliage which completely conceals the natural spacing of the tree crowns. The structure of the forest on a river bank is always different from that in the interior.

Attempts have been made to study the stratification of the Rain forest by plotting the frequency of trees in arbitrarily delimited height classes. Thus Booberg [6] plotted the heights of all the numbered trees in the forest reserves of Java (using the measurements of Koorders); he found that the result gave a continuous curve, with no indication of modes of frequency at certain heights. From this he concluded that the several *étages* (stories) described by authors such as Diels and Rübel cannot be recognised in the forests of Java. A similar curve (Fig. 1.1), but showing a slight tendency to maximum frequency at a certain height class,[1] was obtained by Davis and Richards [7], using measurements of trees on clear-felled plots in Mixed Rain forest in British Guiana. That continuous curves do not disprove the existence of stratification is shown by Vaughan and

[1] The 76–85 ft (23–27 m.) class; this corresponds approximately with the B story (see below).

Wiehe [8], who found that in the 'Upland Climax forest' of Mauritius the frequency of trees plotted against size (diameter) classes gave a continuous curve, though a very marked stratification was readily apparent to an observer (illustrated in their paper by an accurate perspective drawing).

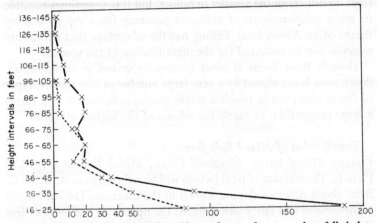

Fig. 1.1   Total heights and heights of lowest leaves of trees on clear-felled plots, Moraballi Creek, British Guiana. From Davis and Richards [7] p. 367. The continuous line represents distribution of total heights among height intervals, the broken line the distribution of the heights of lowest leaves for all trees over 15 ft (4·6 m.) on a total area of 400 × 50 ft (122 × 15·3 m.).

The Profile diagram

Because the direct observation of the stratification of the Rain forest usually offers insuperable difficulties, Davis and Richards [7] adopted the device of constructing profile diagrams to scale from accurate measurements of the position, height and width and depth of crown of all the trees on narrow sample strips of forest. This technique, first applied to the forest of British Guiana, has since been used by various workers in several parts of the tropics and has proved a valuable means of studying and comparing the structure of tropical forest communities. The method may be briefly described as follows:

A narrow rectangular strip of forest is marked out with cords, the right-angles being obtained with the help of a prismatic compass. In Rain forest the length of the strip should not usually be less than 200 ft (61 m.); 25 ft (7·6 m.) has proved a satisfactory width. All small undergrowth and trees less than an arbitrarily chosen lower limit of height are cleared away. The positions of the remaining trees are then mapped and their diameters noted. The total height, height to first (large) branch, lower limit of crown and width of crown of each tree are then measured. Often it is only possible to obtain these measurements by felling all the trees on the strip, and the trees must be felled in a carefully selected order, so that the heavier trees do not crush the smaller in falling, but it is sometimes possible to make measurements of sufficient accuracy from the ground by means of an Abney level. Felling has the advantage that herbarium material can be collected for the identification of the species.

Though Rain forest is most commonly mixed in composition, dominance being shared by a very large number of tree species, types of forest also exist in which a single species is dominant and forms a large proportion, or rarely the whole, of the highest stratum.

Stratification of Mixed Rain forest

Primary Mixed forest, Moraballi Creek, British Guiana (Fig. 1.2; Plate 1). Three strata of trees (which will be referred to as A, B and C, from above downwards) are shown in the diagram. The lowest (C) is continuous and fairly well defined; the upper two are more or less discontinuous and not clearly separated vertically from each other and in the original description (Davis and Richards, [7] pp. 362–72) these were regarded as a single stratum of irregular profile. Most of the gaps in the highest stratum (A) are closed by trees in stratum B; strata A and B together thus form a complete canopy.

The average height of the trees in A on the diagram is about 35 m. but elsewhere usually higher than this (to about 42 m.), that of B about 20 m., while stratum C includes trees between 20 m. and the arbitrary lower limit of 4·6 m. (15 ft), average height about 14 m. On the profile strip (135 ft = 41 m. long) there are sixty-six trees over 15 ft high; seven of these can be reckoned as belonging to the first stratum, twelve to the second and the remainder to the third.

The crowns of stratum A are only here and there in contact laterally, but it must be remembered that its canopy is more closed than it appears to be in the diagram because trees whose crowns overlap the sample strip, but whose bases are outside, are not shown. The trees of A belong to many species and families (Lecythidaceae, Lauraceae, Araliaceae, etc.). Their crowns are usually wider than deep and tend to be umbrella-shaped.

Stratum B is more continuous, but has occasional gaps. Like A it is composed of many species belonging to numerous families

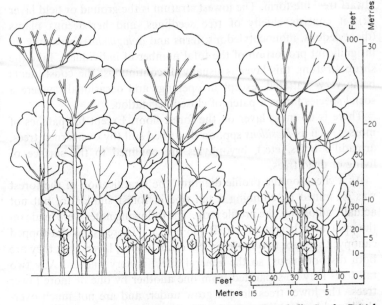

Fig. 1.2 Profile diagram of primary Mixed forest, Moraballi Creek, British Guiana. From Davis and Richards [7] p. 368. The diagram represents a strip of forest 135 ft (41 m.) long and 25 ft (7·6 m.) wide. Only trees over 15 ft (4·6 m.) are shown.

(mostly different from those of A). A considerable proportion of the trees are young individuals of species which reach the stratum A when mature. The crowns tend to be deeper than wide.

In stratum C there are very few gaps, and the density of foliage and branches is greater than at any other level of the forest, higher or lower. More than half the total number of individuals are young trees of species properly belonging to the higher strata; the remainder are small species peculiar to stratum C and mostly belong to families scarcely represented in strata A and B (especially Annonaceae and Violaceae). Both the young A and B trees and the true C species usually have long, tapering, conical crowns, much deeper than wide.

Below the three stories of trees represented on the profile diagram there are two other strata consisting chiefly or partly of woody plants; both are ill defined. The upper of these, the average height of which is about 1 m., may be called the shrub stratum (D), though the constituents include many young trees, small palms, tall herbs (Marantaceae, etc.) and large ferns, as well as small woody plants of shrub or 'dwarf tree' life-form. The lowest stratum is the ground or field layer (E); it consists chiefly of tree seedlings, and herbaceous plants (dicotyledons, monocotyledons, ferns and *Selaginella*) form only an insignificant proportion of the total number of individuals. Like the shrub stratum, this layer is usually discontinuous, the constituents being very scattered, except in openings and occasionally where a social species forms a patch of closed vegetation.

There is no moss layer on the forest floor. Except for patches of mosses such as *Fissidens* spp. on disturbed soil (by overturned trees, armadillo holes, etc.), bryophytes are confined to the surface of living or dead trees.

As a check on the profile diagram, the stratification of the forest was examined by climbing a tall tree in forest similar to, but not actually adjoining, the profile strip. The following notes were made on the spot at a height of 110 ft (33·6 m.): 'There is no flat-topped canopy; there are two more or less clear [upper] layers, but they are both discontinuous, so the general effect is very uneven. Any two tall trees may be separated from one another by one or more lower trees. The lower trees do not grow under, and are not much over-shadowed by the higher. Practically all the lower trees of the canopy are covered and bound together with lianes. [These do not usually reach the trees of the highest stratum.] Apart from the ordinary upper canopy [i.e. stratum A] trees, there are rare ones which tower far above all others. Two such outstanding trees were seen whose whole crowns were well clear of all surrounding trees.' No outstanding trees are shown in the profile diagram; such trees are probably not more frequent than 1 per sq. km. on the average, and their height is about 40–45 m. They belong to species rarely found in stratum A (e.g. *Hymenaea courbaril, Peltogyne pubescens*).

The results of direct observation from 110 ft thus agree closely with the profile diagram. Both methods demonstrate that the forest is three-storied; the rare 'outstanding trees' do not constitute an independent stratum.

Primary Mixed Dipterocarp forest, Mt Dulit, Borneo. The original description of this forest structure is given by Richards [9]. Here again three tree strata are recognisable, but instead of C being definite and A and B ill defined, the highest stratum, though discontinuous, is well defined and sharply separated vertically from B, the division between B and C being difficult to draw. The average heights of the strata are about 35, 18 and 8 m. respectively. On the sample strip there are seven trees in stratum A and eighty-six in B and C (about forty-eight in B and thirty-eight in C); these numbers are not directly comparable with those for the Guiana profile, because the strip was longer (200 ft = 61 m.) and because only trees 25 ft (7·6 m.) high and over (instead of 15 ft and over) were measured.

The crowns of A trees are mostly fairly well separated from each other and raised well above the very dense second stratum; hence, when the forest roof is looked at from above (e.g. from the crest of an escarpment), the individual crowns of the trees in A can be separately distinguished at a great distance. The majority of the trees in stratum A belong to the Dipterocarpaceae. The crowns are about 6 m. deep on the average and tend to be wider than deep. Stratum B is almost completely continuous, each crown usually being in lateral contact with several others; its canopy is almost as dense as that of C. The species here are very numerous and belong to a great variety of families; some immature dipterocarps are of course present, but the majority belong to other families. The crowns are about 4·5–6 m. deep and are mostly less wide than deep.

Stratum C is also continuous. It consists both of young individuals of species reaching strata A or B when mature and of small species which rarely exceed 15m. in height; the former predominate. In both groups of species the crown tends to be very deep and narrow and conical in shape. The small species characteristic of this layer belong to many families; but it is interesting that, as in the Guiana forest, Annonaceae are abundant. The space between strata B and C is filled with trees of intermediate height. Observation over the whole district shows, however, that there are two quite separate groups of species, B species which average about 18 m. when mature and which do not flower when much smaller, and C species, about 8 m. high when mature, which begin to flower when very small. The absence of vertical discontinuity between the middle and lowest tree stories is therefore due partly to the abundance of immature B trees and partly to the presence of individuals of C species taller than the average.

Below the C tree stratum the vegetation becomes much less dense. No very clear stratification is evident in the smaller undergrowth, but it is convenient to divide it into a shrub stratum averaging 4 m. high, consisting of 'shrubs', palms and young trees, and a ground layer up to 1–2 m. high, consisting of tree seedlings and herbs, including ferns,

Selaginella and rarely one or two species of very small palms. The number of species of herbaceous plants is considerable, but the tree seedlings outnumber them in individuals; on ten quadrats 1 m. square the total number of tree seedlings and other woody plants was 184 and of herbaceous plants 135 (shoots). The density of the ground layer is uneven; over large areas it is represented only by widely scattered plants; here and there it is fairly dense, though seldom as dense as in a deciduous wood in Europe.

There is no moss stratum on the ground in the Mixed Dipterocarp forest at low altitudes, but mosses are sometimes found growing on the ground on very steep slopes.

Primary[1] *Mixed forest ('Wet Evergreen forest'), Shasha Forest Reserve, Nigeria.* The structure of this is described in detail by Richards [10]. Like the forest of Guiana and Borneo, the Nigerian forest has three tree strata. Only the lowest story (C), of trees up to 15 m. high (average about 10 m.), is continuous. Above it there is an irregular mass of trees of various heights, the tallest of which is 46 m. high. The crowns of these taller trees are sometimes in contact laterally, but there is no closed canopy above stratum C. Observation over the whole area of which the sample strip formed a part showed that this irregular mass could be separated into two strata, A of trees 37–46 m. (average about 42 m.), and B of trees 15–37 m. (average about 27 m.) high respectively.

The first stratum thus consists of very large trees scattered through the forest (though much more densely than the 'outstanding trees' of Guiana). The crowns of these trees are umbrella-shaped and extremely heavy; they are up to 25 m. or more wide. They are rarely in lateral contact and are raised well above those of stratum B. Story A consists of comparatively few species (in the area studied chiefly *Lophira procera* and *Erythrophleum ivorense*).

Stratum B is also open and the crowns are only occasionally in contact. In the diagram the gap in this layer beneath the crown of the

---

[1] Work by the author (1948, unpublished) makes it seem likely that this forest has suffered disturbance in the past and is old secondary rather than truly primary; it is, however, probably sufficiently mature to differ little in structure from the primary (climax) forest of the region. The same is probably true of nearly all so-called virgin or primary forest in Nigeria, and perhaps in the whole of West Africa. Forest which has never at any time been cultivated exists in West Africa on swampy sites (Fresh-water Swamp forest), but elsewhere only on extremely limited areas, mostly on steep rocky slopes, etc.

large stratum A tree is noteworthy. The trees of stratum B have small narrow crowns, usually under 10 m. wide. The species are numerous and belong to a wide range of families.

Stratum C is very dense and almost without gaps. The crowns are packed closely together and are usually tightly bound together by lianes. The majority of the trees in this layer are species which never reach a higher story; they belong to various families, but there is a tendency for a single species (the actual species varying from place to place) to be locally dominant. The remaining trees in the stratum are chiefly young B species; young A species seem to be strikingly rare. The C species, like those of the corresponding layer in Guiana and Borneo, mostly have small conical crowns, but old individuals sometimes have remarkably wide and heavy crowns.

The shrub stratum (D) is very indefinite. It consists largely of young trees belonging to strata B and C, so there is no clear division between this layer and the lowest tree story. Species properly belonging to stratum D (most of which are 'dwarf trees' rather than true shrubs) are few. The density of the shrub stratum is very variable; in undisturbed forest it is never so dense as to make progress difficult, and in some places both shrubs and ground layers are almost wanting.

The lowest layer of the forest is the ground layer (E), which consists of plants varying from a few centimetres to 1 m. or more high. The components are tree seedlings, dicotyledonous and monocotyledonous herbs, and ferns, the first generally predominating. This stratum is even more unevenly developed than D. Large stretches of the forest floor may be almost completely bare, but in places, especially in openings, the ground may be concealed by a dense growth of herbaceous plants and tree seedlings. There are no mosses on the ground.

A photograph taken at 78 ft (24 m.) above ground in forest similar to the sample strip from which the profile was drawn is reproduced on Plate 2. This shows the dense mass of small trees with interwoven crowns about 9–12 m. high. Taller trees rise above this compact layer to various heights, but they do not form a closed canopy at any level, so that above 11 m. it is possible to see clearly for some distance in any direction. When climbing this tree one seemed to emerge into full daylight as soon as the dense C stratum was passed. The picture of the stratification obtained from this tree is thus very similar to that disclosed by the profile diagram.

The great discontinuity of the upper two tree strata is a remarkable feature of this forest. A very open stratum is perhaps a consequence of the relatively severe seasonal drought, and may be a general feature of West African evergreen forest, since Aubréville [11] describes the 'Closed forest' of the Ivory Coast as consisting of a dense mass of vegetation 20–30 m. high dominated by scattered taller trees; he compares its structure to a *taillis-sous-futaie* (coppice-with-standards). It is also noteworthy that in the 'Evergreen Seasonal forest' (*Carapa–Eschweilera* association) of Trinidad [12, 13], the canopy of the highest tree stratum is very discontinuous. On the other hand, since the forests of the Shasha Reserve have been much modified by selective felling and native cultivation, it is possible that the openness of the upper strata is due to removal of a proportion of the larger trees, or that the community studied is an advanced stage in a secondary succession and not in fact a true primary forest.

The three examples of Mixed Rain forest which have just been described are thus similar in the main features of their stratification though they show some differences which may be of significance. From these data and from the other information available the following general statements about the structure of Mixed Rain forest seem justified:

(i) There are five strata of independent plants (in the sense already given): the three tree layers, which we have termed the A, B and C strata respectively,[1] consisting entirely of trees, a D layer consisting mainly of woody, but often partly of herbaceous, species, which it is convenient to call the shrub stratum, though only a few of its components are shrubs in any exact sense of the word, and lastly the ground or field layer (E) of herbs and tree seedlings.

(ii) The height of each stratum varies from place to place, but not within wide limits. Thus the height of stratum A is about 30 m. or more in the Guiana forest described above, about 35 m. in the Borneo example and about 42 m. in the Nigerian example. Similarly, the height of the B strata is 20, 18 and 27 m. respectively, and the C 14, 8 and 10 m. respectively in the three examples.

(iii) Stratum A usually has a more or less discontinuous canopy, though there is considerable variation in this respect between the three profiles described; possibly in the tallest and most luxuriant

[1] To avoid confusion with the ecological use of 'dominant' and 'subdominant', these terms are not applied here to strata, as is the common practice in forestry literature.

Rain forest (e.g. in parts of the Malay Peninsula where the average height of the highest tree story is said to exceed 200 ft (61 m.)), this layer may be practically continuous. There is some evidence that the A stratum becomes increasingly discontinuous as the climatic limits of rain forest are approached. The B stratum may be continuous or more or less discontinuous; C is always more or less continuous and is often the densest layer of the forest.

(iv) A vertical discontinuity between the canopies of neighbouring strata may or may not be apparent. Thus in the Guiana forest there is some discontinuity between the B and C strata, but little between A and B; in the Bornean forest there is a conspicuous gap between the canopies of A and B, but none between B and C; in Nigeria, as in Guiana, the main discontinuity is between B and C. The vertical limits of the shrub and ground layers are never very clearly defined.

(v) Each stratum in the forest has a different and characteristic floristic composition, but in all the strata except A and B young individuals of species which reach higher strata when mature form a large proportion of the total number of individuals.

(vi) The trees of each stratum have a characteristic shape of crown. In A the crowns tend to be wide or even umbrella-shaped, in B they are as deep as wide, or deeper, in C conical and tapering, much deeper than wide. In each stratum the trees also have other characteristic physiognomic features which are not a function of their taxonomic affinities.

REFERENCES

[1] GAMS, H. (1918) 'Prinzipienfragen der Vegetationsforschung', *Vjschr. naturf. Ges. Zürich*, LXIII 293–493.

[2] SAXTON, W. T. (1924) 'Phases of vegetation under monsoon conditions', *J. Ecol.*, XII 1–38.

[3] BROWN, W. H. (1919) *Vegetation of Philippine Mountains* (Manila).

[4] MILDBRAED, J. (1922) *Wissenschaftliche Ergebnisse der zweiten deutschen Zentral-Afrika-Expedition 1910–1911 unter Führung Adolf Friedrichs, Herzog zu Mecklenburg* (Leipzig).

[5] CHEVALIER, A. (1917) 'La forêt et les bois du Gabon', *Les Végétaux utiles d'Afrique tropicale française*, fasc. 9 (Paris).

[6] BOOBERG, G. (1932) 'Grondvormen, étages, en phytocoenosen van Java's vegetatie', *Hand. 6 de ned.-ind. natuurw. Congr., 1931*, 329–46.

[7] DAVIS, T. A. W., and RICHARDS, P. W. (1933–4) 'The vegetation of Moraballi Creek, British Guiana: an ecological study of a limited area of Tropical Rain Forest. Parts I and II', *J. Ecol.* XXI 350–84; XXII 106–55.

[8] VAUGHAN, R. E., and WIEHE, P. O. (1941) 'Studies on the vegetation of Mauritius: III. The structure and development of the Upland Climax Forest', *J. Ecol.*, XXIX 127–60.

[9] RICHARDS, P. W. (1936) 'Ecological observations on the rain forest of Mount Dulit, Sarawak. Parts I and II', *J. Ecol.*, xxiv 1–37, 340–60.

[10] —— (1939) 'Ecological studies on the rain forest of southern Nigeria: I. The structure and floristic composition of the primary forest', *J. Ecol.*, xxvii 1–61.

[11] AUBRÉVILLE, A. (1933) 'La forêt de la Côte d'Ivoire', *Bull. Com. Afr. occid. franç.*, xv 205–61.

[12] BEARD, J. S. (1944) 'Climax vegetation in tropical America', *Ecology*, xxv 127–58.

[13] —— (1946) 'The natural vegetation of Trinidad', *Oxf. For. Mem.*, no. 20.

2 Regeneration Patterns in the Closed Forest of Ivory Coast

A. AUBRÉVILLE

Translated by S. R. Eyre from 'La Forêt Coloniale', *Académie des Sciences Coloniales: Annales*, IX (Société d'Editions Géographiques, Maritimes et Coloniales, Paris, 1938) 126–37.

DESCRIPTIVE STUDY OF THE COMMUNITIES

A DEFINITE kind of order exists in the closed forest in spite of the diversity of species. Each locality has certain dominant or abundant species, but a statement regarding their relative numbers is only valid for a restricted area. They change very quickly from place to place. In a neighbouring locality a rise in the numbers of new dominants will be apparent while those of the first area are scarcely represented. This is because the gregarious species, with fairly rare exceptions, are not distributed continuously throughout their range. They are distributed throughout the forest as are the spots on the skin of a panther – in groups, outside which they are thinly scattered. So marked is this that, when surveying along a track, one may fail to see certain dominant species of the region because they are grouped outside the particular line being followed. Nevertheless, the longer the line of transect, the greater the chance of observing all the characteristic species of the region. Conversely, if a patch in the form of a long narrow band (a fairly common phenomenon) is traversed longitudinally by the surveyor, he will be tempted to conclude that the species which *appears* to him to be widespread and abundant in the area as a whole is, in fact, growing only within a few feet of the path he has followed.

Quite frequently the species which are abundant or dominant in the forest are fairly few in number. It is because of this that certain phytogeographers, such as Chipp [1], have applied the association concept to the equatorial forest; this is a little surprising in view of the long list of species which compete here for light and space. Moreover, as pointed out by Gaussen [2], the term 'association' is not very appropriate: 'It is not implied that the plants provide each other with any protection whatsoever. Generally speaking they are competitors.' The usual groupings cannot be regarded as having only one or two dominant and abundant species, as a number of cases cited below will show. Within the same floristic region it is impossible, in every locality, to identify, as Chipp has done, two species which characterise the community. In fact the equatorial forest 'association' (to use the convenient term but the one which leads to confusion) is much more complex than this. It is made up of any combination of the gregarious species which are characteristic of a formation. It is possible that one or two of these species appear more frequently as dominants than do the others. It does not follow, however, that they alone characterise the community since they may not be found over wide areas covered by the same formation. But let us return to the rather over-simplified associations of Chipp. One might think that in the formation normally regarded as being coterminous with the *Lophira–Cynometra* association, for example, one is bound to find these two abundant or dominant species everywhere in association with each other. There is, in fact, little likelihood of this. The same association appears in Ivory Coast but does not have any more importance than other associations. It is inevitably associated with 'rain forests' certainly, just as are the *Lophira–Entandrophragma* association, the *Tarrietia–Anopyxis* association and many others; in the same way *Triplochiton–Piptadenia* must be associated with 'deciduous forests'.

In the various parts of the rain forest in the Man region in Ivory Coast, taking only the dominant species, one is able to distinguish the following associations:

1. *Alstonia congensis*
   Bussea occidentalis
   Carapa procera
   Chidlovia sanguinea

2. *Lophira procera*
   Uapaca guineensis
   Chidlovia sanguinea
   Coula edulis
   Carapa procera

3. *Petersia africana*
   Parinarium kerstingii
   Piptadenia africana
   Calpocalyx brevibracteatus

4. *Turreanthus africana*
   Petersia africana
   Chidlovia sanguinea
   Coula edulis

5. *Tarrietia utilis*
   Lophira procera
   Turreanthus africana
   Coula edulis
   Chidlovia sanguinea

6. *Piptadenia africana*
   Tarrietia utilis
   Funtumia africana

7. *Piptadenia africana*
   Ceiba pentandra
   Copaifera ehie
   Petersia africana
   Chidlovia sanguinea
   Bussea occidentalis

8. *Petersia africana*
   Piptadenia africana
   Tarrietia utilis
   Triplochiton scleroxylon
   Bussea occidentalis
   Chidlovia sanguinea

9. *Petersia africana*
   Chidlovia sanguinea

10. *Lophira procera*
    Petersia africana
    Tarrietia utilis
    Piptadenia africana
    Placodiscus boya
    Ochtocosmus africanus, etc.

11. *Petersia africana*
    Piptadenia africana
    Funtumia africana

12. *Petersia africana*
    Chlorophora excelsa
    Chidlovia sanguinea

The following species often recur as dominants: *Lophira procera, Chidlovia sanguinea, Petersia africana, Bussea occidentalis, Tarrietia utilis, Piptadenia africana, Turreanthus africana, Coula edulis, Calpocalyx brevibracteatus* – nine in all. At each observation point there is a different combination of several of them along with one or two other species.

Moreover, to give a more exact impression of the average composition of the plant population, it is necessary to add the following species which are often quite abundant: *Uapaca guineensis, Parkia bicolor, Anopyxis occidentalis, Erythrophleum ivorense, Funtumia africana, Parinarium kerstingii, Octoknema borealis*.

All these species are characteristic of the 'rain forests'. Consequently, the citing of just one association, dominated by two species chosen from amongst the most extensive ones, to represent the Man area, would provide only a very arbitrary picture.

What is said below regarding the Man example could equally well be said about the survey we have made in the different regions of the colony. In the 'rain forests' in the hinterland of Tabou there appear, apart from some of the species already listed, some other dominant ones, such as *Cynometra ananta, Strombosia pustulata, Scytopetalum tieghemii, Pachylobus deliciosus, Dalium aubrevillei, Protomegabaria stapfiana, Calpocalyx aubrevillei, Parkia bicolor, Monopetathus* sp., *Diospyros sanza minika*.

In fact one can form the following impression of the closed forest. Some of the environmental conditions which hold sway over a vast area bring into existence a certain type of forest made up of a large number of species having the same biological requirements. At each point in the forest a particular set of these species dominates, but the composition varies spatially. Out of the total number of species, twenty or so become dominant more frequently than the rest. These twenty, most particularly, characterise the formation.

A census carried out over an area of 210 ha. in the Banco forest reserve in coastal 'rain forest' confirms this point. 5703 trees with a diameter of at least 0·50 m. were counted in this forest management operation. The main species were as follows:

Dominant species		per cent
'Avodiré'[1]	*Turreanthus africana*	18
Red ironwood	*Lophira procera*	10
'Adjouaba'	*Pachylobus deliciosa*	10
Stinkwood tree	*Petersia africana*	7

Fairly abundant species		
'Lo'	*Parkia bicolor*	4·8
'Dabéma'	*Piptadenia africana*	4·5
Oil-bean tree	*Pentaclethra macrophylla*	3·7
Akee apple	*Phialodiscus bancoensis*	3·3
'Melegba'	*Berlinia acuminata*	3·1
African walnut	*Coula edulis*	3·0
'Aramon'	*Parinarium kerstingii*	2·4
'Poé'	*Strombosia pustulata*	2·3
'Bodioa'	*Anopyxis occidentalis*	2·2
Guinea plum	*Parinarium tenuifolium*	2·1

Thus *four dominant species* comprised about 45 per cent of the large tree population, *ten abundant species* formed 33 per cent, and *fifty scattered species* accounted for no more than 22 per cent.

DYNAMIC STUDY OF THE CLOSED FOREST

After man has destroyed the primeval forest in order to cultivate the humus-rich soil, a new forest community soon takes possession of it. The appearance and floristic content of the vegetation are modified. The new community is 'secondary forest', so called by M. Chevalier in contradistinction to the pre-existing virgin forest, which is termed 'primary'. Certain phytogeographers do not approve of these terms,

---

[1] Common names in quotation marks are the original French renderings of native names presented by the author. Where possible English equivalents from F. R. Irvine's *Woody Plants of Ghana* (1961) are given here [Ed.].

doubtless because, for them, the flora and the physiognomy of a plant community are constantly modified by diverse influences until a final stage in stable equilibrium is reached, with maximum bio-logical production preordained for each locality, which they call 'the climax'. Thus, before a plant community reaches its final state, it will pass through numerous intermediate stages. Under the climatic conditions of lower Ivory Coast, the so-called 'secondary' forest is one of these unstable stages, but it is not necessarily the 'second' one. As for the forest communities which appear to be 'primary', they are not necessarily final stages or climaxes. Nevertheless, for practical purposes, and seeing that we do not have any more meaningful expressions, these terms may be accepted as having some meaning. The closed primary forest is a formation of large trees, normally having a hard wood. The total canopy is very dense, direct sunlight does not reach the ground, and a man can walk about freely. Secondary forest originates under conditions which are well under-stood. It is composed of a particular set of species and generally it changes back only slowly to the primary forest type. When the secondary species are fairly large, they provide only a thin cover (as in the case of the umbrella tree, *Musanga smithii*) beneath which the seeds of primary forest species germinate successfully. Thus, at length, the primary forest is re-established. Doubtless its floristic composition will no longer be that of the pre-existing one destroyed by man, but the physiognomy will become the same as that of a 'primary' forest.

One could reserve the name 'primary forest' for the forest which has never been modified by man. The true virgin forest doubtless exists in Ivory Coast within generally inhabited regions, although it is never possible to be certain that any particular area has not actually been worked over centuries ago. Thus there is no recognisable difference between forests which are incontestably virgin since remote times, and *ancient* forests which are found in regions which were worked over by man a long time ago. They have the same species and the same associations of species.

It is only the forests which have re-formed recently which have a special floristic composition. A number of large trees still survive which are of secondary origin such as the black bark (*Terminalia ivorensis*), 'iroko' (*Chlorophora excelsa*), 'fraké' (*Terminalia altissima*), oil palm (*Elaeis guineensis*), 'Bahé' (*Fagara macrophylla*), silk cotton tree (*Ceiba pentandra*), 'ouochi' (*Albizia zygia*), 'eho' (*Ricinodendron africanum*), 'loloti' (*Lannea acidissima*) and African tragacanth (*Sterculia tragacantha*). One must not infer from this that each time one finds these species in closed forest, this is evidence for former felling and burning: in the 'rain forests', they are able to invade into gaps which occur sporadically due, for example, to the falling of large trees. In the 'deciduous forests', moreover, they are often true climax species. The oil palm is an exception: it never occurs in virgin forest. If it is found, this is evidence of former occupation by man. In fact one never finds young oil palms beneath the cover of old forest.

It is very easy to study the successive stages in woodland re-establishment on ground which has been cleared by indigenous cultivators and then abandoned to the wild. It is always the same species which invade first, the chief ones being as follows. The umbrella tree (*Musanga smithii*) forms pure, uniform stands. Representatives of the Euphorbiaceae are numerous, there being several spiny species of *Macaranga*, the 'lié' (*Phyllanthus africanum*), the beard tree (*Tetrorchidium didymostemon*) and the 'tougbi' (*Bridelia micrantha*). The Leguminosae are represented by two species of *Albizia*, the 'bangbaye' and the 'ouochi' (*Albizia sassa* and *A. zygia*). Some other abundant species are as follows:

Ulmaceae:	'adaschia' (*Trema guineensis*)
Hypericaceae:	'ouombé' (*Haronga paniculata*)
Compositae:	'iaonvi-poupouia' (two species of *Vernonia*)
Loganiaceae:	cabbage palm (*Anthocleista nobilis*) and *Gaertnera paniculata*
Verbenaceae:	several species of *Vitex* and *Premna hispida*
Rutaceae:	several species of *Fagara*
Apocynaceae:	'déchavi' (*Rauwolfia vomitoria*), some species of *Conopharyngia* and the two species of *Funtumia* (*F. elastica* and *F. africana*)
Rubiaceae:	some *Canthium*, *Grumilia*, etc.
Moraceae:	some *Ficus*, African bread fruit (*Treculia africana* and some *Myrianthus*
Myristicaceae:	African nutmeg (*Pycanthus kombo*)
Anacardiaceae:	'loloti' (*Lannea acidissima*)
Combretaceae:	black bark (*Terminalia ivorensis*)

Subsequently, other species make their appearance in the shade of the first ones. They vary very much from place to place, coming as they do from neighbouring communities.

It is more difficult to study changes over a period of time in the different forest successions which make up an old forest of the 'primary' type. The species of secondary scrub have a rapid rate of growth; in some months, in some years, one can observe the development of this scrub very well, but the situation is different in the high forest. Chipp has initiated investigations into this kind of situation in the forest of the Gold Coast [1]. He has recognised the *Lophira–Cynometra* as a final optimum stage or 'climax' in the thick forest. Neighbouring stages he has correlated with slightly different environmental conditions and regarded as pre-climaxes: *Lophira–Entandrophragma*, *Entandrophragma–Khaya* and *Triplochiton–Piptadenia*. According to him the *Anopyxis–Tarrietia* and *Chlorophora–Landolphia* associations only represent temporary stages, though they are sufficiently stable to persist for a very long period.

The point already made in the foregoing descriptive study of communities regarding these 'climax' associations can be repeated here. They certainly correspond very closely to apparently stable forms in the forest, but in our opinion this view of the facts is over-simplified since a large number of other combinations of species could equally well be regarded as the climax. The climax or pre-climax communities *Lophira–Entandrophragma* and *Anopyxis–Tarrietia* are to us merely selected examples, more theoretical than real, from the climax of the 'rain forest' formation. Similarly the climax associations *Triplochiton–Piptadenia* and *Chlorophora–Landolphia* are selected examples from the corresponding climax 'deciduous forest'. The pre-climax *Entandrophragma–Khaya* may represent transition types between 'rain forests' and 'deciduous forests'.

In real terms, all those groupings of dominant species from which examples have been cited with regard to our surveys in the Man region are climaxes. The 'rain forest' or the 'deciduous forest' are final forms in stable equilibrium with environmental conditions and they possess very numerous climaxes (any combination whatsoever out of a large number of abundant species in the formation).

Moreover, is it not agreed that the term 'climax' should be applied strictly to those groupings of dominant species which characterise 'synecies' in the closed forest (to use the vocabulary of the phyto-sociologist)? The communities under discussion must be final, stable 'synecies', unless it can be demonstrated that they tend to become more and more homogeneous and ultimately to be reduced to only a few climax types (as Chipp's thesis would suggest). In fact they do not persist in the same condition, and it is possible that *none of them* can be considered to be the last rung of the ladder.

Not only do the synecies vary spatially (in the ways that we have shown), they also do so in time without ever culminating in a definite climax. Before presenting some facts which support our hypothesis, it is relevant to note that if, within a virgin forest of considerable extent but with identical environmental conditions (one can still find this despite the fact that today the remains of it are more and more exploited by the natives), the 'synecies' develop towards a single climax, it seems likely that one ought to be able to demonstrate a certain homogeneity of population. In fact none is to be found: the population make-up varies just as capriciously as in forests where one has reason to suspect the former influence of man – just as much as one might expect in forests which were still seeking their final point of equilibrium.

Let us consider a few facts. One can frequently find in the primary forests (possibly virgin) that there is a great difference between the species-composition of the adult tree population and that of the developing one which is destined to replace it. Often an abundant species in the community is not regenerating in the undergrowth. The natives express this situation by saying that these trees 'never make little ones'. Sometimes in fact our surveyors had difficulty in finding young individuals of a species in order to fell them; we only encountered large individuals from which it was impossible to collect herbarium specimens. I remember in particular in the vicinity of Abengourou, having encountered for the first time a community of a very large tree with a very hard wood – the asamela (*Afrormosia elata*) – we searched for a long time, without any success, for a young tree of this species so that we could cut it down easily. We were forced to resign ourselves to felling a large tree. The natives of that place maintained that the tree 'did not make little ones'. On the other hand one finds areas invaded by seedlings and young plants of species of large trees which are quite absent from the upper layers of the forest.

We have also the example from the forest reserve of Massa Mé where we formerly carried out a census over an area of about 1·4 ha. in the interior of untouched primeval forest, on a stony, laterite soil. The forest is very old, the canopy dense, the soil unmodified and there are few lianes. Above all, the highest stratum is composed of very thick, widely-spaced trees – the criterion for a very ancient forest.

All the trees having a diameter of 5 cm. or more were counted, as well as all the young individuals belonging to some selected species. The inventory is shown in Table 2.1.

The species-composition on this area, which was taken at random in the forest, is very varied, since seventy-four different species were found in it. It is immediately noticeable how the species at the head of the list, which form the upper stratum, are poorly represented in the lowest stratum and in the undergrowth. Among the very large trees, the 'dabéma' (*Piptadenia africana*) predominates. The impression which this species gives in reality is more striking than might be thought from a mere reading of the figures, for the vast, flattened crowns of the 'dabémas', with their large aerial spread and conspicuous buttresses, lead surveyors to believe that they have found a considerable population of them. Nevertheless, in the future, the 'dabéma' will almost disappear from this patch of land since there are only two saplings for replacement. The stinkwood tree (*Petersia africana*) will doubtless hold its own but the following very large species will disappear completely: African elemi (*Canarium sweinfurthii*), 'lo' (*Parkia bicolor*) and Guinea plum (*Parinarium tenuifolium*).

What will the future population of large trees be, if we imagine that all the present saplings of 10 to 20 cm. diameter will one day be competing for the light? In Table 2.1, all the names of species which are capable of attaining very large size are given in capitals. It is obvious that the struggle will be severe. If certain species now occupying the forest canopy are about to decline, a large number of others are ready to replace them. At the end of the list it can be seen that very large species are represented, each by a single pole. Consequently they must have established themselves here quite recently. These species are: the African mammy apple (*Mammea africana*), the red ironwood (*Lophira procera*), the bark-cloth tree (*Antiaris welwitschii*), the sasswood tree (*Erythrophleum ivorense*) and the 'bodioa' (*Anopyxis occidentalis*). It is impossible to have any idea as to which will be the future dominants, but it is clear that plenty of combinations are possible and that the advance towards a climax (if such a thing can be visualised for equatorial formations) is a very indirect one.

The inventory of young plants of the abundant species is equally interesting, since it is established that species such as the 'niangon' (*Tarrietia utilis*), the scented guarea (*Guarea cedrata*) and the 'avodiré'

TABLE 2.1

Systematic name	Family	Seedlings	Metres											
			0·1	0·2	0·3	0·4	0·5	0·6	0·7	0·8	0·9	1·0	1·2	1·5
PIPTADENIA AFRICANA	Leg.	—	—	1	—	—	1	1	—	1	—	1	—	2
PETERSIA AFRICANA	Leg.	—	4	3	2	—	1	2	1	1	—	1	1	—
KHAYA IVORENSIS	Mel.	12	1	—	—	—	—	—	—	—	—	—	—	—
CANARIUM SWEINFURTHII	Bur.	—	1	—	—	—	—	1	1	—	—	—	—	—
KLAINEDOXA GABONENSIS	Irv.	—	—	—	—	—	1	1	—	—	—	—	1	—
PARKIA BICOLOR	Leg.	—	—	—	—	—	1	1	—	1	1	—	—	—
PARINARIUM TENUIFOLIUM	Ros.	—	—	—	—	—	—	1	—	—	—	—	—	—
Pachylobus deliciosa	Bur.	—	46	37	11	—	—	1	—	—	—	—	—	—
Funtumia africana	Apoc.	—	3	2	2	—	1	—	—	1	—	—	—	—
Diospyros sanza minika	Eb.	—	45	27	2	2	—	1	—	—	—	—	—	—
STROMBOSIA PUSTULATA	Olac.	—	43	13	4	2	—	—	—	—	—	—	—	—
SCOTTELIA CORIACEA	Flac.	—	15	7	3	2	—	—	—	—	—	—	—	—
Monodora myristica	Anon.	—	3	6	3	2	—	—	—	—	—	—	—	—
PARINARIUM KERSTINGII	Ros.	—	2	—	2	2	2	—	—	—	—	—	—	—
HANNOA KLAINEANA	Sim.	—	16	6	3	—	—	—	—	—	—	—	—	—
Allanblackia parviflora	Gutt.	—	15	3	3	1	—	—	—	—	—	—	—	—
Scythopetalum tieghemii	Scyt.	—	3	3	1	4	—	—	—	—	—	—	—	—
Calpocalyx brevibracteatus	Leg.	—	16	5	2	—	2	—	—	—	—	—	—	—
Cola maclaudii	Sterc.	—	3	6	1	—	—	—	—	—	—	—	—	—
DANIELLIA Aff. THURIFERA	Leg.	—	3	2	1	2	1	—	—	—	—	—	—	—
Protomegabaria stapfiana	Euph.	—	5	3	1	—	—	—	—	—	—	—	—	—
Trichoscypha arborea	Anac.	—	2	5	—	2	—	—	—	—	—	—	—	—
Panda oleosa	Pand.	—	6	3	2	—	—	—	—	—	—	—	—	—
Cola nitida	Sterc.	—	6	5	1	—	—	—	—	—	—	—	—	—
IRVINGIA GABONENSIS	Irv.	—	2	3	1	—	—	—	—	—	—	—	—	—

Species	Family							
ENTANDROPHRAGMA								
ANGOLENSE	Mel.	3	3	2	2	–	–	–
Diospyros kamerunensis	Eb.	–	22	4	–	–	–	–
GUAREA CEDRATA	Mel.	7	7	2	1	–	–	–
TARRIETIA UTILIS	Sterc.	53	6	2	1	–	–	–
AMPHIMAS PTEROCARPOIDES	Leg.	–	6	2	1	–	–	–
PHIALODISCUS								
PLURIJUGATUS	Sapi.	–	2	2	–	1	–	–
Baphia pubescens	Leg.	–	10	3	–	–	–	–
Enantia polycarpa	Anon.	–	3	3	–	–	–	–
Isolona campanulata	Anon.	–	5	3	–	–	–	–
Aporrhiza rugosa	Sapi.	–	3	3	–	–	–	–
ENTANDROPHRAGMA								
CYLINDRICUM	Mel.	1	–	2	–	–	–	–
TURREANTHUS AFRICANA	Mel.	48	4	–	1	1	1	–
AFZELIA BELLA	Leg.		3	–	2	2		
Albizzia zygia	Leg.		1	2	–	1		
Trichilia thompsonii	Mel.		1	1	1			
Coula edulis	Olac.		1	2				
Pleiocarpa mutica	Apoc.		7	2				
Vitex micrantha	Verb.		3	2				
Garcinia polyantha	Gutt.		4	2				
MACROLOBIUM								
CHRYSOPHYLLOIDES	Leg.		3	1	1			
Discoglypremna caloneura	Euph.		5	1				
Ricinodendron africanum	Euph.		–	1				
Tylostemon mannii	Laur.		2	1				
Trichilia heudelotii	Mel.		2	1				

TABLE 2·1—cont.

Systematic name	Family	Seedlings	Metres				
			0·1	0·2	0·3	0·4	0·5
Lannea acidissima	Anac.		–		–	–	1
Octoknema borealis	Oct.		–	1	–		
Xylopia elliotii	Anon.		3	1			
Pycnanthus kombo	Myr.		4	–		1	
Cola mirabilis	Sterc.		37	1	1		
Myrianthus arboreus	Mor.		12	–			
Conopharyngia durissima	Apoc.		8	1			
Phyllanthus discoideus	Euph.		2	1			
Bridelia micrantha	Euph.		–	1			
Napoleona sp.	Lecy.		9				
Omphalocarpum sp.	Sapi.		5				
Fagara macrophylla	Rut.		4				
Uapaca guineensis	Euph.		3				
MAMMEA AFRICANA	Gutt.		1				
LOPHIRA PROCERA (ALATA)	Och.		1				
ANTIARIS WELWITSCHII	Mor.		1				
ERYTHROPHLEUM IVORENSE	Leg.		1				
ANOPYXIS OCCIDENTALIS	Rhiz.		1				
Anthocleistha nobilis	Log.		1				
Scaphopetalum amoenum	Sterc.		5				
Microdesmis puberula	Euph.		3				
Baphia nitida	Leg.		3				
Carapa procera	Mel.		1				
Randia genipaeflora	Rub.		1				
CHLOROPHORA EXCELSA	Mor.	1	–				

(*Turreanthus africana*), although they have no seed parents either in the area concerned or in the immediate vicinity, seem to invade the undergrowth. Their arrival is recent.

Such examples of detailed surveys are interesting but, unfortunately, rare. In order to draw up such an inventory of the forest, it is necessary to have at one's disposal several excellent native surveyors. Such surveys would demonstrate, without doubt, that there is not always a close correspondence between the actual population of dominant species and that which is waiting in the undergrowth to replace it. They would direct attention to the confusion that exists between the static view of the plant association in the equatorial forest and the dynamic one of the climax. In the area that we have surveyed, the traveller sees, as we have said, a community characterised by *Piptadenia africana*. This is a false impression: the only species which are really in possession of the soil and which appear to have a firm hold are found in the lower levels of the forest. Here these are the 'adjouaba' (*Pachylobus deliciosa*), the flint bark (*Diospyros sanza minika*), the 'poé' (*Strombosia pustulata*), the odoko (*Scottelia coriacea*) and perhaps the stinkwood tree (*Petersia africana*).

Chipp has cited the red ironwood (*Lophira procera*) as a typical species of the climax in these forests of warm and very humid lands – formations which have a dense canopy and which epitomise the maximum development of a vegetation community. Now we have often observed – without, unfortunately, being able to make a detailed count – that beneath very old high forest in which the red ironwood dominated, one found few young plants of this species. There is nothing surprising in this if one knows about the light-demanding propensities of *Lophira procera*. It makes only a sorry rate of growth beneath dense shade, whereas it grows luxuriantly and rapidly in full sunlight. The red ironwood invades secondary scrub, roadsides and river banks without difficulty, indeed everywhere where it finds a little sunlight. Its fruits are winged and its colonising ability is great. But how could such a species be a typical element in a climax of dense, dark 'rain forest'?

Again, let us compare two species of large tree as seen from a census taken from management fellings in 80 ha. of primary forest in the Banco reserve (Table 2.2). In the series, it is quite normal to find a progressive decrease in the numbers of African walnut (*Lovea klaineana*) as diameter increases. The small number of large trees,

TABLE 2.2

	Metres													
	0·05	0·10	0·20	0·30	0·40	0·50	0·60	0·70	0·80	0·90	1·00	1·10	1·20	1·30
African walnut (Lovea klaineana)	199	47	28	12	8	5	4	3	1	1				
'Dabéma' (Piptadenia africana)	3	16	27	32	38	45	44	25	27	24	21	6	6	1

those which have passed the sapling stage, indicates that this species only established itself in the community a short time ago and that it is tending to occupy a more and more important position. On the other hand, one cannot but be astonished at the large number of old 'dabémas' (*Piptadenia africana*) in comparison to the scarcity of young trees. Is one not compelled to recognise on the one hand a rapidly expanding species and, on the other, one which is now dominant but which is going to decline?

M. Brillet, Inspecteur des Forêts in Indo-China, has cited [3] the very interesting case of the 'lim' (*Erythrophleum fordii*), a beautiful species in Tonkin. In the forests where old 'lims' are abundant one would find only very few seedlings of this species around seed parents – in most cases none at all. On the other hand, elsewhere one would find pure stands of 'lim' composed of *poles* of the same age.

REFERENCES

[1] Chipp, T. F. (1927) *The Gold Coast Forest: A Study in Synecology.*
[2] Gaussen, Henri (1933) *Géographie des Plantes.*
[3] Brillet, M. (1927) *Les Annales Forestières de l'Indochine.*

3 The Seasonal Forests of Trinidad

J. S. BEARD

From 'The natural vegetation of Trinidad', *Oxford Forestry Memoirs*, no. 20
(Clarendon Press, Oxford, 1945) pp. 83-7, 91-4, 100-6.

SEMI-EVERGREEN SEASONAL FOREST

Distribution

THIS formation (Plate 3) contains two associations: the *Peltogyne porphyrocardia* (purpleheart) and the *Trichilia smithii–Brosimum alicastrum* (acurel–moussara) with two and three faciations respectively.

Peltogyne	(i) *Protium–Tabebuia* faciation (incense–poui). Mountains of the north coast and north-west peninsula.
	(ii) *Mouriri* faciation (bois lissette). Moruga district, Guayaguayare, Manzanilla, Toco.
Trichilia–Brosimum	(iii) *Bravaisia* faciation (jiggerwood). South coast Erin to Marac, Manzanilla, Bois Neuf.
	(iv) *Protium* faciation (gommier). South coast Moruga to Guayaguayare, Mayaro.
	(v) *Ficus* faciation (figuier). Limestone mountains at Tamana, Brigand Hill and Cumaca.

Structure

This is a somewhat open type of forest (see Fig. 3.1). Canopy is generally formed between 20 and 40 ft by a lower story, above which stands a layer of scattered emergent trees attaining as a rule heights of 60–80 ft, though heights of over 100 ft have been recorded. The gregariousness of this upper story is very variable. In the *Peltogyne* association the emergents habitually grow so thickly as almost to form a continuous upper canopy: in the *Trichilia–Brosimum* association they do so occasionally, by groups.

The number of individual trees over 1 ft girth per acre varies from 115 to 175. Sizes attained at maturity differ rather more widely in the various types, as the following table shows:

	No. of individuals per 100 acres		
	Over 1 ft girth	Over 6 ft girth	Over 10 ft girth
Peltogyne:			
Protium–Tabebuia	11,500	350	20
Peltogyne: Mouriri	17,600	544	92
Trichilia–Brosimum:			
Bravaisia	16,700	360	123
Protium	12,500	123	18
Ficus	13,300	1,042	417

Mature trees do not normally attain more than 5 or 6 ft girth, though they often run to enormous sizes in the *Trichilia–Brosimum–Ficus* type. The tremendous girths of old trees are in fact a prominent feature of the latter, and girths of 30 ft have been recorded.

The number of large-stemmed trees per acre is unconnected with the gregariouness of the upper story. The *Protium–Tabebuia* faciation has relatively few stems of considerable girth but possesses a thick upper story, this being composed mainly of trees in the 4–5 ft girth class.

Trees almost never show a clear stem of over 40 ft. They fork or branch low down and 20–30 ft is the common stem-length. In the

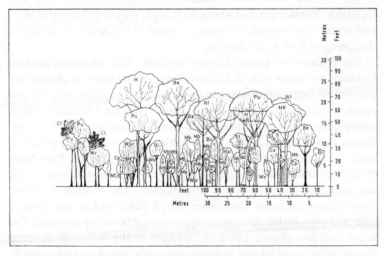

Fig. 3.1 Profile diagram of Semi-evergreen Seasonal Forest measured in the Rochard Douglas Reserve

lower story the clumped growth of the *Coccoloba* species is a feature. Crowns in the lower story vary with exposure, being conical when suppressed but flattened when given full light. Crowns of the emergents tend to be wide-spreading and somewhat umbrella-shaped, supported on heavy branches.

Physiognomy and life-form

Lianes here reach their optimum development in Trinidad. All large trees are commonly loaded with vines all over the crown, so much so that in some cases they seem to cause breakage of limbs. Large ropes hang from the branches, and even the lower story may be infested so that passage is difficult without a cutlass. No floristic studies in the lianes have been undertaken.

Epiphytes are typically only moderately to poorly developed. Aroids and orchids are relatively rare; bromeliads commonly form colonies on the branches of all the larger dominant trees, though not on the trunks or on the smaller trees. As described by Broadway and Smith [2], they exhibit no local peculiarities and the same species of *Tillandsia* and *Vriesia* may be looked for as in the *Carapa* forests.

There is a noticeable absence of ferns and mosses from this formation. The only native bamboo, *Guadua latifolia* Lam., is to be found in the *Trichilia–Brosimum* association, though it is very rare and localised. A small bamboo about 20 ft high, growing in small clumps and armed with unpleasant spines, it inhabits the banks of ravines in the Quinam–Palo Seco districts.

The position of palms is very variable. The *Protium–Tabebuia* faciation possesses only a *Coccothrinax* sp. (latanier), a denizen of the ground layer. Similarly, the *Ficus* faciation features only *Bactris cuesa* (grigri) and *Geonoma vaga* (anaré). The other faciations show strong development of palms. The *Bravaisia* type contains a *Sabal* sp. (carat) as the commonest species in the lower story at the rate of some 20 to the acre: the *Coccothrinax* appears in the Palo Seco district, coastally or in the neighbourhood of mud-volcanoes, and *Bactris major* (roseau) occurs gregariously. The *Protium* faciation shows about the same abundance of the *Sabal* and in the *Mouriri* type it is even better developed, approaching 50–60 to the acre. The latter faciation shows, however, a phase in the Windbelt Reserve on the east coast where *Scheelea osmantha* (trash palm) replaces *Sabal*, and a further phase on the eastern foothills of the Northern

Range at Toco which is without palms. It would appear as though the large palms avoid mountainous country.

Buttressing, with a few exceptions, is but seldom present. Large trees tend instead to flute slightly at the base. The exceptions are formed by *Ficus tobagensis*, whose large plank buttresses are noticeable, by *Peltogyne porphyrocardia*, which has moderate buttresses, and by old trees of *Ceiba occidentalis*, *Cedrela mexicana* and some other dominants, which may develop buttresses. Large trees are noticeably shallow-rooting and frequently show large sprawling exposed roots.

Two species only possess stilt roots, *Bravaisia integerrima* (jiggerwood) and *Virola surinamensis* (cajuca). The latter is very rare in this formation and is confined to occasional moist places. *Bravaisia* breaks the rule associating stilt roots with swamp habitat and shows no particular site preference: it may be found on the driest sites.

Thorniness is not an abundant character. All are unarmed except for six species of the upper story and three of the lower: these, however, are save one, definitely characteristic species of this formation. They are *Hura crepitans* (sandbox), *Chlorophora tinctoria* (fustic), *Fagara martinicensis* (l'epinet), *Fagara trinitensis* (bosoo), *Machaerium robinifolium* (saltfishwood), *Ceiba occidentalis* (silk cotton), *Bumelia grisebachii* (acoma piquant), *Erythrina pallida* (wild immortelle) and *Xylosma seemanni* (wild cerise).

Cauliflory is throughout absent as a character of the upper story, though present in 4–12 per cent of the species of the lower story, representing 8–33 per cent of the individuals.

Degree of deciduousness and the leaf characters are shown in the following table, where all types of deciduous species are included under that head.

These figures may be compared with those for evergreen seasonal forest.

Pinnate arrangement is predominant among the compound leaves. Leaflets range from 2 to 50, the average being 7 to 10.

The degree of deciduousness varies considerably throughout; the lower story is more evergreen than the upper, which is to be expected as the lower story is sheltered from the wind, thus encountering lower evaporating power of the air at critical drought periods. The deciduousness of the lower story parallels variations in that of the upper fairly closely in most cases. The percentage of deciduous species in the upper story varies from 26·0 to 42·3, the average being about 33·3 (one-third). The deciduous species are not among the most numerous in individuals, so that the percentages of deciduous individuals are less high, ranging from 10·8 to 29·7, with the average at about 16·6 per cent (one-sixth).

Compound leaves predominate both in point of species and individuals in the upper story, and simple leaves in the lower story. Leaves of the upper story are overwhelmingly mesophyllous, while those of the lower story are less markedly so, showing some tendency to be microphyllous.

The type of evergreen leaf is dark green and shiny above, often hard and leathery, with a drip-tip. The flush-leaf is bright red or pale green. The type of deciduous leaf is a mid-green above and not shiny, soft and easily wilting: drip-tips are often present. The flush-leaf is pale green.

Shrub and ground layers. A ground layer is virtually absent and the soil bare. The thick carpet of dead leaves formed by the annual leaf-fall very swiftly decomposes and for the bulk of the year there is little visible humus. The shrub layer is, however, very strongly developed. Where the upper story is tending to form canopy, the shrub layer may thin out or be composed largely of young palms. Habitually, however, it is a dense growth merging up into the lower forest story and composed of hard woody little trees, mainly myr-taceous – *Eugenia, Myrcia* and *Calyptranthes* spp. – except where in the *Trichilia–Brosimum* association clumps of the gregarious *Bactris major* occur. There is a most noticeable scarcity of herbaceous growth in these strata.

Reproduction

Very much the same remarks apply here as to evergreen seasonal forest, in that the bulk of the species have large heavy seeds relying on animals and birds for dispersal. The following species have winged seeds: *Tabebuia* spp., *Terminalia amazonia, Cedrela mexicana* and *Ceiba occidentalis.* The fruit of *Hura crepitans* is a capsule about 3 in. in diameter which bursts violently in the dry season and scatters the seeds. There do not appear to be any other special dispersal mechanisms. The greater part of the most common species have fruits with an edible pulp which are very attractive to parrots.

	Percentage of species							Percentage of individuals						
	Deciduous	Compound leaves	Raunkiaer's leaf classes					Deciduous	Compound leaves	Raunkiaer's leaf classes				
			Lepto-phyll	Nano-phyll	Micro-phyll	Meso-phyll	Macro-phyll			Lepto-phyll	Nano-phyll	Micro-phyll	Meso-phyll	Macro-phyll
1. Peltogyne–Protium–Tabebuia														
Upper story	42·3	61·6	..	..	19	69	12	29·7	56·4	..	..	20	77	3
Lower story	20·8	16·7	..	..	12	67	21	10·1	3·4	..	..	37	51	12
2. Peltogyne–Mouriri														
Upper story	35·7	57·0	..	2	10	86	2	16·6	54·4	..	½	9	90	½
Lower story	12·5	25·0	3	..	16	66	15	1·2	20·6	½	..	25½	73	1
3. Trichilia–Brosimum–Bravaisia														
Upper story	37·2	58·1	..	2	9	89	..	18·4	53·3	..	1	6	93	..
Lower story	12·0	20·0	..	..	8	80	12	0·8	33·2	..	..	15	71	14
4. Trichilia–Brosimum–Protium														
Upper story	30·3	60·6	..	..	6	94	..	10·8	67·5	..	..	1	99	..
Lower story	Nil	31·2	6	..	25	69	..	Nil	39·9	3	..	14	83	..
5. Trichilia–Brosimum–Ficus														
Upper story	26·0	48·0	..	..	4	87	9	20·9	41·1	..	..	1	92	7
Lower story	9·4	18·8	..	..	16	75	9	10·4	44·2	..	..	3	89	8

Habitat

This formation occurs in certain areas where a more severe drought is encountered during the dry season than in the evergreen seasonal forest areas. The drought is not invariably set up by lack of rainfall, but is due to a combination of the factors of topography, exposure, rainfall and soil-type, as the following notes will show.

(i) *Peltogyne* association. *Protium–Tabebuia* faciation (purple-heart–incense–poui type).

> *Topography*. Mountainous and exposed to sea winds.
> *Local climate*. Coastal to moist seasonal.
> *Soil type*. Free-draining loam over schist.
> *Human interference*. Apparently absent.

This faciation is confined to the lower slopes of the mountains facing the sea along the north-west coast. The rainfall is adequate for evergreen forest, but the habitat is desiccated by the steep topography and sea winds so that a semi-evergreen type only can develop.

(ii) *Peltogyne* association. *Mouriri* faciation (purpleheart–bois lissette type).

> *Topography*. Dissected peneplain.
> *Local climate*. Moist seasonal.
> *Soil type*. Red clays with impeded drainage.
> *Human interference*. Apparently absent.

This faciation reaches its most widespread development in the Moruga district, but is also found at various points close behind the windward littoral from Toco to Guayaguayare wherever red clay soils are found. It appears that, although these areas are near to the sea, they are seldom much affected by exposure to sea winds, but rather through low rainfall, since it appears that immediately along the windward seaboard the rainfall is rather lower than it is at short distances inland. In this case the dry habitat is set up mainly by the low rainfall, aggravated by the impermeable nature of the soil which absorbs but little water and dries out hard with deep cracks in the dry weather. The rainfall, being relatively low in total amount compared to the rainfall of evergreen forest areas, is correspondingly relatively low during the dry season, so that relatively more intense drought is felt. The connection of this faciation with red-weathered clay soils, showing a bright red colour at the surface, is most noticeable: it does not occur on brown clays. The red colour appears to indicate a soil whose internal drainage conditions are extremely bad, and which is highly deficient in lime and humus. The latter deficiency causes the soil to be compact and impermeable.

(iii) *Trichilia–Brosimum* association (acurel–moussara). *Bravaisia* faciation (jiggerwood type).

Topography. Dissected peneplain.

Local climate. Moist seasonal.

Soil type. Marls, brown or yellow non-calcareous clays with impeded drainage.

Human interference. Active.

This faciation is typically developed along the 'southern range' from Erin to Marac, and occurs also at North Manzanilla and on hillocks in the Nariva swamp. Nearness to the littoral in these cases is probably in the main a matter of lowered rainfall, but exposure to sea winds has undoubtedly a certain effect. The effect of relatively low rainfall is aggravated by this exposure, by the topography and by the hard, impermeable nature of the soil. Soil conditions, however, are not as adverse as with the red clays of the *Peltogyne–Mouriri* type. The brown and yellow clays, while low in lime, are rich in humus, and frequently the soil is marly with excellent lime and humus status, so that permeability and water conduction are improved. The frequently calcareous nature of the soil probably explains the abundance of *Brosimum alicastrum* and the former abundance of *Cedrela mexicana* (now reduced by logging), these being two species which reach their optimum development on limestones.

There has been much biotic interference with this type, but it appears that the effect has been much less than formerly supposed. The discovery of kitchen middens along the coast has led to the supposition of widespread shifting cultivation by aboriginals in pre-Columbian times, but there is no actual evidence. This theory is supported by observations in British Honduras that 'the more recent secondary growths contain *Cedrela mexicana* (Cedar) and *Brosimum alicastrum* (breadnut) which is typical of secondary forest on calcareous soils' [3], and that *Brosimum alicastrum* occurs on and around Maya ruins [1]. Further, it offers a convenient method of explaining the origin of the *Brosimum–Melicocca* society: both these species having edible fruit, stands may have sprung up around encampments. *Mora excelsa* has undoubtedly been assisted to spread in this way. Of recent years, fires have frequently occurred in the forest and there has been active logging for *Cedrela* and *Chlorophora*. In 1933 the forest was very badly damaged by a cyclonic storm; the enumeration surveys on which the present description of the association is based date, however, from before this.

In the writer's opinion human interference has undoubtedly had considerable effect on the floristic composition, particularly at the western end, where the variation in dominance is most marked. Such variation is usually a sign of second-growth conditions (cf. associes of deciduous seasonal forest), and in this case it would be difficult to account for it satisfactorily otherwise. The human interference responsible for this, however, is likely to have been in the form of fires of fairly recent years and not connected with aboriginal cultivations. The structure of the forest has not been markedly affected and is in general agreement with that of the adjoining *Mouriri* and *Protium* types, where for reasons of topography and position there is unlikely to have been ancient interference. The occurrence of semi-evergreen forest here is quite satisfactorily explained by reference to natural factors, and desiccation of the habitat by interference is not a necessary presumption. One feels it unlikely that the aborigines chose this area for shifting cultivation since the lack of rainfall would imperil the crops; there has been experience of this of recent years with establishment of teak plantations in this area under a 'taungya' system with the aid of field crops. The dense growth of *Brosimum* and *Melicocca* on ridges in the east is paralleled by the behaviour of *Peltogyne* and *Mouriri* in their association, and does not appear to need any explanation connected with edible fruit. It would be difficult, further, to see why these two species should be specially selected for their fruit when the island contains several better species – *Manilkara* for example. *Brosimum* and *Cedrela* reach, in Trinidad, their optimum development on limestones but in virgin forest: their behaviour is therefore different here from what it is in British Honduras, and the abundance of these species in this association is more likely to be due to calcareous soil than to second-growth conditions.

(iv) *Trichilia–Brosimum* association. *Protium* faciation (acurel–gommier type).

Topography. Hilly, often moderately exposed to sea-winds.

Local climate. Moist seasonal.

Soil type. Free-draining sands over sandstone.

Human interference. Negligible.

A dry habitat is set up by combination of low rainfall with steep topography and freely draining sandy soil. Moisture penetrates but quickly drains away. The faciation is principally distinguished from the *Bravaisia* by substitution of arenophilous species for calcicole. It is doubtful if there has been much human interference as most of the area is inaccessible.

(v) *Trichilia–Brosimum* association. *Ficus* faciation (moussara–figuier type).

Topography. Mountainous, with very steep slopes. Altitude 600–1000 ft.

Local climate. Wet seasonal.

Soil type. Clay over limestone, usually very thin: excessive subsoil drainage.

Human interference. Appararently absent.

Here rainfall is high and the dry season not marked, but a dry habitat is set up by the excessive subsoil drainage of the porous limestone. At critical drought periods this substantially reduces available moisture and the effect is the same as with a low rainfall on a more normal soil. Owing to the inaccessible nature of the area, human interference has not been active, but wind damage has had something of the same effect. Calcicole species are prominent.

DECIDUOUS SEASONAL FOREST

Distribution

Deciduous seasonal forest in Trinidad (Plate 4) appears originally to have been represented by a faciation of the *Bursera–Lonchocarpus* association which occurs in Tobago, the Lesser Antilles and Venezuela. This forest has suffered very severely from human interference and no undisturbed examples can be found. As it exists at present the forest must be described as a facies – the *Machaerium robinifolium* facies – of the association. We may also include with deciduous seasonal forest for descriptive purposes another community which will be called the *Protium–Tabebuia* ecotone and seems to have been degraded by fires from the *Protium–Tabebuia* faciation of the *Peltogyne* association of semi-evergreen seasonal forest. The former community is found on the islands off the north-western peninsula, on Pointe Gourde, and the lower slopes of the peninsula itself; the latter on the foothills and southern slopes of the western Northern Range.

Structure

The canopy is a low one, being formed by a low story at 10 to 30 ft. Above this layer stand out irregularly scattered emergent trees which may attain 60 ft in height. Even the lower story is by no means continuous, so that the canopy is not entirely closed (see Fig. 3.2).

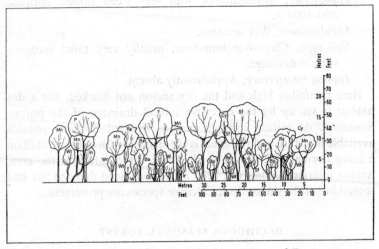

Fig. 3.2   Profile diagram of Deciduous Seasonal Forest measured at Chaguaramas

There are few very large trees. The emergents may attain up to 8 ft in girth, though this is rare, 5 ft being the usual upper limit. Species of the lower story do not exceed 3 ft. The average number of individuals per 100 acres over 1 ft girth is about 15,000, over 6 ft girth 45, and over 10 ft nil.

Trees of the lower story have conical crowns when dominated, but pyramidal or flattened crowns when exposed. The emergent trees have rounded crowns with a tendency towards flattening. Stems fork or branch low down and tend to be bent or crooked.

Physiognomy and life-form

Presumably owing to the dryness of the habitat, lianes and epiphytes are not over-well developed here. The former are about as abundant as in evergreen seasonal forest, whereas the latter are almost absent. A number of aroids and bromeliads normally epiphytic have, however, become terrestrial, particularly on rocky places. The same is true of the strangling epiphyte *Clusia rosea* (matapal).

Palms are very rare, the solitary representative being *Maximiliana elegans* (cocorite). Trees are not buttressed, and stilt roots are absent. Ferns and any other special forms are absent, though succulents appear in a littoral society. The vegetative characteristics are as follows:

(a) By number of species
  Emergents
    Evergreen 57·6%   Deciduous 42·4%
    Compound leaves 57·6%   Average No. of leaflets 9·9
    Macrophyllous 9·1%, Mesophyllous 72·7%,
                            Microphyllous 15·2%
  Lower Story
    Evergreen 88·9%   Deciduous 11·1%
    Compound leaves 26·7%   Average No. of leaflets 7·5
    Macrophyllous 13·4%, Mesophyllous 75·5%,
                            Microphyllous 11·1%
(b) By number of individuals
  Emergents
    Evergreen 35·9%   Deciduous 64·1%
    Compound leaves 68·3%
    Macrophyllous 11·7%, Mesophyllous 43·1%,
                            Microphyllous 45%
  Lower Story
    Evergreen 74·2%   Deciduous 25·8%
    Compound leaves 26·5%
    Macrophyllous 12%, Mesophyllous 57·1%,
                            Microphyllous 30·9%

The community is thus predominantly deciduous, rather more markedly in the upper than in the lower story, due to the shelter from sun and wind which the lower story receives. The emergents are tending towards smallness of leaf, a high proportion being microphyllous. Compound leaves predominate over simple. The predominant evergreen leaf-type is microphyllous, dark green, shiny and leathery with a drip-tip. The deciduous leaf-type is mesophyllous, pale green, often tough and seldom shiny above, soft and limp rather than leathery. The flush leaves are bright green.

The deciduous period is the dry season, January to April. Leaves fall in January and the new flush commonly appears in April *before* the break of the rains.

The shrub layer merges with the lower story of the forest.

Ground vegetation is remarkably scarce. Except for an occasional seedling, herb or tuft of coarse grass the soil is quite bare often even of dead leaves, as the annual leaf-fall decomposes rapidly in the ensuing rainy season. Terrestrial bromeliads flourish chiefly on rocks, but *Bromelia karatas* (pinguin) is occasional throughout.

Reproduction

Most of the constituent species have, like those of the other forests, heavy seeds with no particular mechanism for dispersal. Notable exceptions are the *Tabebuias* and *Terminalia amazonia* which have light winged seeds. As usual birds are probably the chief distributing agents. Many of the species, notably *Machaerium* and *Lonchocarpus*, appear to owe their present abundance in the face of human interference to their coppicing power which regenerates them freely after cutting.

Cacti often propagate vegetatively by layering.

Habitat

(i) *Machaerium* facies.

Topography. Mountainous.

Climate. Rainfall 45 to 60 in. per annum with dry season January to April.

Soil type. Free-draining loam over schist or limestone.

Biotic interference. Very active.

The habitat is a dry one, the rainfall being low and aggravated by the steep topography and freely draining soil. The dry-season drought is intense, the five dry-season months frequently having less than 2 in. of rain each. The dryness is further aggravated by human interference, which by destruction of the forest canopy lets sun and wind in to get at the soil. Human activity in this area is of long standing. In Spanish times much of it was under cultivation for sea-island cotton. There is still sporadic shifting cultivation of maize and other food crops and irregular culling for firewood. There is, however, no grazing and fires are rare, probably owing to the lack of herbaceous ground vegetation. After felling for firewood or shifting cultivation there is little or no invasion of vines and second-growth weeds (*Cecropia, Ochroma, Vismia* spp.) such as occurs in moister areas. The forest therefore grows back rapidly, to a large extent from coppice.

(ii) *Protium–Tabebuia* ecotone.

Topography. Mountainous, mainly of south aspect.

Climate. Rainfall 60 to 80 in. per annum with dry season January to April.

Soil type. Free-draining loam over schist.

Biotic interference. Very active.

The habitat, while not quite so dry as that of the preceding associes, remains of considerable dryness, the rainfall being fairly low and aggravated by the steep topography with its aspect mainly south, free-draining soil and human interference. The area has been under shifting cultivation for many years and valuable trees have been felled for timber. Felled sections become quickly loaded with vines so that regrowth is retarded. Further, gardens are always burned before planting and fire often escapes into the surrounding forest. In years of severe drought, which seem to occur about every fourteen years, widespread damage is done by fires.

REFERENCES

[1] BARTLETT, H. H. (1936) 'A method of procedure for field-work in tropical American phytogeography', *Botany of the Maya Area* (Carnegie Inst., Washington).

[2] BROADWAY, W. E., and SMITH, L. B. (1933) 'The Bromeliaceae of Trinidad and Tobago', *Proc. Amer. Acad. Arts and Sci.*, LXVIII.

[3] HARDY, F., SMART, H. P., and RODRIGUEZ, G. (1935) *Some Soil-types of British Honduras, Central America*, Studies in W.I. Soils, IX (Trinidad) 56 pp.

4 An Example of Sudan Zone Vegetation in Nigeria

R. W. J. KEAY

From 'An example of Sudan zone vegetation in Nigeria', *J. Ecol.*, XXXVII (1949)
337–40, 344–8, 353, 360–1.

DETAILS OF THE ZURMI AREA

Position

THE Zamfara Forest Reserve is in the north-eastern part of Sokoto
Province and lies between latitudes 12° and 13° N. and on longitude
7° E. (see Fig. 4.1). It has an area of about 240,000 ha. and is roughly
rectangular in shape. The eastern boundary of the reserve is the
boundary between Sokoto and Katsina Provinces; another large
reserve in Katsina is contiguous with it.

The ecological work was done in an area of roughly 3000 ha. on
either side of the river Fafara between Zurmi and Dumburum, only
24 km. south of the international boundary.

Climate

The nearest station for which adequate meteorological data are
available is Katsina, about 90 km to the east of Zurmi and 13 minutes
of latitude farther north. The recorded mean annual rainfall is about
710 mm. for a period of twenty-three years, but was slightly higher in
the period 1925–35, to which the data in Table 4.1 refer.

Isohyetal maps (see Fig. 4.2) show that the rainfall in the Zurmi
area must be much the same as that of Katsina. The climate is
typical of stations in the central part of the Sudan zone.

Geology and topography

The whole area is occupied by granites and gneisses of the ancient
basement complex, which forms the central massif of northern
Nigeria. It is probable that during an arid phase of the Quaternary a
mantle of sand (mostly of local origin) was laid over the irregular
surface of the crystalline rocks. This mantle of sand, sometimes very
thick, sometimes quite thin, makes the present general land surface
more or less level. The sandy mantle is frequently pierced by rock
outcrops, either because it was insufficient to cover them or because

TABLE 4·1

Month	Mean rainfall		Temperature (°C)		Relative humidity	
	Amount in mm.	No. of days	Mean max.	Mean min.	Mean 9 a.m.	Mean 3 p.m.
January	0·3	0	30·0	13·0	22	18
February	Nil	0	32·7	15·2	20	16
March	0·5	0	38·0	19·0	25	15
April	6·3	1	40·2	22·5	33	17
May	61·2	6	38·1	23·5	56	32
June	90·7	7	35·2	21·1	68	48
July	151·4	12	31·7	20·6	79	68
August	281·2	17	30·0	20·9	86	72
September	129·0	11	31·4	21·3	81	65
October	9·6	1	34·0	20·0	58	30
November	0·3	0	35·0	16·9	26	16
December	Nil	0	31·0	13·8	28	14
Year	730·5	55	33·9	19·0	49	34

of subsequent denudation. Some outcrops are merely surface exposures, while others are 'whale-back' hills, or heaps of boulders, up to 40 m. high. A hill called Dutsen Bagai, and two others near it, are composed of very hard black boulders of garnetiferous quartzite and a basic rock of the Charnockite group.

In the valley of the Fafara river (see Fig. 4.3) steep rocky slopes separate the sandy alluvium of the river terraces from the general high-level ground of the surrounding country. A number of small tributaries join the Fafara inside the Reserve. They are mostly quite short and are completely dry in the dry season; many of them commence in semi-basin sites, and enter the main stream in steep-sided gulleys.

History

Information has been obtained from the writings of Clapperton and Denham [1], Barth [2], Hogben [3] and Niven [4]. Clapperton passed through Zurmi in May 1824, and it is clear from his account that the surrounding country was fully occupied and had been so for a century or more. These were troublous times, however, owing to the Fulani rising which started at Sokoto in 1804. Barth, who visited Zurmi in 1853, noted that Zamfara 'is at present divided into petty states each of which follows a different policy'. Some towns adhered to the original rulers, while others supported the Fulani, but no matter

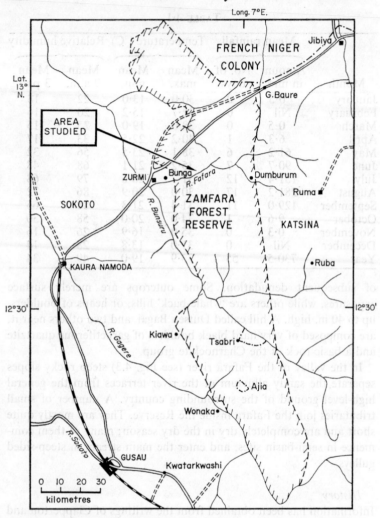

Fig. 4.1 Map showing Zamfara Forest Reserve and the area studied

what its allegiance any small town or village was liable to be ruth-
lessly raided by its enemies. Through this kind of fighting the country
which is now Forest Reserve became uninhabited, the people either
fleeing to the large towns, or else being slaughtered, or carried off
as slaves, while their farmlands became covered with comparatively
dense savanna vegetation.

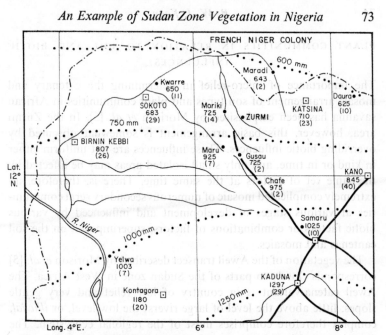

Fig. 4.2 Isohyetal map of north-western Nigeria. (Mean annual rainfall in mm. is shown beneath each station, the figure in parentheses referring to the number of years on which the mean is based.)

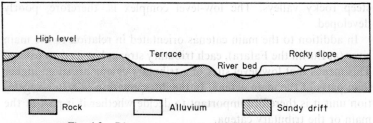

Fig. 4.3 Diagrammatic section of the Fafara valley

As far as one can gather, raiding and punitive expeditions continued in Zamfara from Barth's time right up to 1902 when the British expedition arrived at Sokoto. During the peace which followed, the tendency in Sokoto, as elsewhere, has been for people to spread from the larger towns into the surrounding country which the Fulani wars had rendered desolate. It appears, however, that the area which is now the Zamfara Forest Reserve has remained more or less uninhabited since Barth's time. In fact, the Reserve was constituted in 1919 because most of the area was uninhabited.

PLANT COMMUNITIES IN RELATION TO SOIL AND BIOTIC INFLUENCES

The importance of micro-relief in determining the catenary and mosaic arrangement of soil types and plant communities in African savanna has been emphasised by Morison *et al.* [5]. In the Zurmi area, however, this basic arrangement is somewhat obscured by overriding biotic influences. These influences are not uniform either in kind or in time, and only small isolated areas may be affected by the same set of factors at the same time. There is, therefore, an extremely complicated mosaic of numerous secondary seral communities at various stages of development and influenced by various biotic factors or combinations of factors, superimposed on the soil cantenas and mosaics.

The vegetation of the Aweil transect described by Morison *et al.* [5] corresponds closely to parts of the Sudan zone in West Africa. The Aweil catena-variant is in country of low relief and very gentle slopes, little above the level of large rivers; the low level, or *illuvial*, complex therefore comprises most of the regional ecosystem. The Zurmi area, although comparable to Aweil floristically, is, however, very different topographically, the bulk of the land being *high-level*, well above the flood level of the rivers which flow through relatively steep rocky valleys. The low-level complex is, therefore, poorly developed.

In addition to the main catenas orientated in relation to the main river (at Zurmi the Fafara), each tributary stream has its own catena. This means that low-level sites of a tributary catena are often higher than parts of the main catena. In the interpretation of a soil–vegetation unit it is therefore important to decide whether it is part of the main or the tributary catena.

The main parts of the catenas in the Zurmi area are:

(*a*) High-level – covering most of the area.

(*b*) Rocky hills and slopes – mosaics of colluvial and eroded soils.

(*c*) Low-level – including low-level portions of the tributary catenas as well as those of the main catena.

Vegetation on high-level sites

The three most abundant trees on high-level sites are *Anogeissus leiocarpa*, *Combretum glutinosum* and *Strychnos spinosa*. *Combretum* and *Strychnos* appear to have similar ecological requirements and together, or separately, they are dominant in communities which avoid the moister soils with a water-retentive clay-rubble layer and tolerate the drier sands. *Anogeissus*, on the other hand, tends to be dominant on the moister soils.

The three species may be mixed, but more usually one or other is dominant. Where *Anogeissus* is abundant *Combretum* and *Strychnos* are either absent or rare, and where the other two species are dominant *Anogeissus* is rare. In four enumeration plots (0·4 ha.) more than 70 per cent of the trees are *Anogeissus*; in three of these four plots *Combretum* and *Strychnos* are absent, and in the fourth they make up only 4 per cent. *Combretum glutinosum* and *Strychnos spinosa* make up over 55 per cent of the trees in five plots, none of which has more than 8 per cent *Anogeissus*.

The mosaic of communities dominated either by *Anogeissus* or by *Combretum–Strychnos* corresponds to some extent to the mosaic of soil types, but the vegetation mosaic is complicated by biotic interference.

Most of the vegetation has grown up on cultivated lands, but cultivation is no longer an active determining factor. Today the important factors are grass fires and cattle. The fire tends to keep the vegetation open, to eliminate certain species, and hence selectively to favour fire-resistant species, especially *Combretum glutinosum* and *Strychnos spinosa*. The effect of cattle is more complex and may work in opposition to the fire factor. Trampling and grazing reduce the grass cover, especially by main cattle routes, and thereby stop grass fires not only on the trampled area but also in untrampled patches surrounded by trampled 'fire-breaks'. This fire protection favours the regeneration of trees, especially of *Anogeissus*, whose early stages are very sensitive to fire; old trees of *Anogeissus* are, however, fairly resistant. The foliage of *Anogeissus* is distasteful to cattle; thus once a thicket of young growth is established it is avoided by the animals.

The abundance of *Anogeissus* by cattle routes is most striking. All stages from young thickets to mature woodland are present. In the early stages *Anogeissus* is mixed with *Dichrostachys glomerata*, but the latter is seral and does not persist.

The complex action of the soil mosaic and the two main biotic factors upon the vegetation is best understood by considering an 'ideal' mesic community in which *Anogeissus* (upper stratum) and *Combretum* and *Strychnos* (lower stratum) are codominant trees. Such a community exists under what appear to be mesic conditions of soil and relatively little biotic disturbance. Under the influence of more extreme edaphic or biotic factors the three dominants segregate, as indicated in the following diagram.

Soil type A [with water-retaining layer]	Mesic conditions of soil and little biotic disturbance	Soil type B [without water-retaining layer]

Segregate I *Anogeissus* open savanna woodland with regeneration thickets		Segregate III *Anogeissus–Combretum –Strychnos* open savanna woodland with *Anogeissus* regeneration thickets

Cattle   *Cattle*

Anogeissus–Combretum– Strychnos open savanna woodland

Fire   *Fire*

Segregate II *Combretum–Strychnos* open savanna; a few large *Anogeissus* but no regeneration thickets		Segregate IV *Combretum–Strychnos* open savanna; *Anogeissus* absent or very rare

This segregation of the dominant species, under the influence of edaphic and biotic factors, appears to be comparable to the *associa- tion-segregation* of the forests of North America described by Braun [6]. The Zamfara communities are certainly not climax and so the term *associes-segregate* is used here.

Segregates I and IV are the most distinct because the two sets of factors work in the same direction. In segregates II and III the factors work in different directions, and the resulting communities are of an intermediate type. The biotic factors do not necessarily remain constant either in kind or intensity, and the vegetation mosaic may be further complicated by the initiation of, say, cattle trampling in segregate II, or of fire in segregate III, when these would tend to change to segregates I and IV respectively.

The foregoing explanation is based only on observations of the vegetation as it is today, the effect of the biotic factors not having been tested in controlled experiments. The biotic factors at present controlling any particular community and the ecological behaviour of the more abundant species are, however, evident in this vegetation which may be regarded as a mosaic of numerous natural experiments. The hypothesis of associes-segregates does, at all events, go a long way in explaining the mosaic which at first sight seems unordered.

Vegetation on rocky slopes and hills

The density of vegetation on rocky slopes and hills depends basically upon the proportion of bare rock, and of eroded and colluvial types in the soil mosaic. Where bare rock is extensive the vegetation consists merely of isolated trees (with herbs and shrubs) in the crevices. Optimum development is found where there is a good proportion of colluvial material amongst the boulders. On steep slopes free from rocks most of the soil is badly eroded, and the vegetation is very poor open savanna.

The most significant features of the vegetation on rocky slopes and hills are: (a) the abundance of deciduous scandent shrubs which form a tangled understory; (b) the scarcity of grass; and (c) the protection that the rocks afford against fire and cattle. These features are interrelated and are of great importance in understanding the ecological status of the vegetation of the whole area.

Present-day farming practice shows that it is unlikely that rocky slopes and hills were ever cultivated, and field observations indicate that the rocks give some protection against fire. In the intensively used land by towns, such sites may be used for grazing goats and for providing fuel, but in more remote country it is very unlikely that either cattle or fuel cutters would bother about the rocky areas. It is therefore evident that such sites have suffered less disturbance than others.

Even small groups of boulders have their distinctive vegetation: one or two large trees (e.g. *Diospyros mespiliformis*, *Tamarindus indica* and *Ficus platyphylla*) draped with scandent shrubs. These clumps of dense vegetation appear as islands in the surrounding savanna.

Similar communities are developed on a larger scale on rocky slopes and hills. Several species are confined to these sites, such as *Boscia salicifolia* and *Isoberlinia tomentosa* (a common species of the Guinea zone); a number of other Guinea zone species (e.g. *Pterocarpus erinaceus* and *Khaya senegalensis*) are found only on rocky hills and by watercourses.

ECOLOGICAL STATUS

In trying to determine the climatic climax of this region it would be desirable to examine undisturbed communities on the high-level site. The high-level site, as Morison *et al.* [5] point out, represents the climatic normal in the sense that the soil receives neither water nor dissolved nor suspended matter from other areas. In the Zurmi area, however, the vegetation of high-level sites is greatly disturbed, having been cultivated up to about ninety years ago and still being influenced by grass fire and cattle. These biotic influences are not uniform throughout the area, either in kind or in time, and the vegetation is consequently an extremely complicated mosaic of numerous secondary seral communities at various stages of development and influenced by various biotic factors or combinations of factors. This biotic complex is superimposed on an almost equally complex mosaic of soil types which may vary considerably in quite small areas. Under these influences the dominant species are segregated into communities of equal rank comparable to Braun's *associes-segregates* [6]. Seral development within each segregate may be recognised (e.g. *Anogeissus–Dichrostachys* thickets develop into *Anogeissus* woodland), but the communities of the whole high-level complex are not arranged in a single sere leading to one climax.

Chevalier [7] recognised the secondary nature of most Sudan zone vegetation, and stated that relics of the primitive vegetation are now found only on uncultivable rocky plateaux and stony hills and along watercourses. In the Zurmi area it is clear that the most undisturbed vegetation is today found on rocky hills and slopes. The river terraces are much disturbed but the small streams probably less so. A streamside is, however, more favourable as regards water supply than the high-level complex, so its vegetation cannot be equivalent. The most likely places to seek vegetation approximating to the climatic climax is, therefore, rocky hills, steep and rocky enough to afford protection from cultivation, fire and cattle, but not so steep and rocky as to be quite different from the normal high-level site.

The features of this vegetation are: (i) *form*, dense savanna woodland; (ii) abundant deciduous scandent shrubs; (iii) grass only local; (iv) relatively numerous woody species; (v) *Anogeissus* not overwhelmingly dominant, several other species (e.g. *Lannea microcarpa*) being frequent in the upper stratum; (vi) *Combretum glutinosum* and

Strychnos spinosa, both frequent but not overwhelmingly dominant in the lower stratum.

Such low rocky hills are naturally a little more favourable than the ordinary high-level site, as the presence of such trees as *Diospyros* and *Khaya* indicates, but it seems that the vegetation does approximate to what may be the climatic climax of the area as a whole.

This hypothesis is supported by a study of vegetation in other parts of Africa with similar climate and soil. Hoyle (verbal communication) informs me that vegetation approximating to that of the rocky hills described above occurs on ordinary high-level sites in the Anglo-Egyptian Sudan in country north of the Aweil area to which reference has already been made [5].

In 1947 I paid a short visit to the Sesheke District of Northern Rhodesia, about 100 km. west of the Victoria Falls. A distinct type of vegetation known as 'mutemwa' occurs there to the south of the *Isoberlinia–Brachystegia* woodlands (similar to the Northern Guinea zone of Nigeria). The climate of this area is similar to that of the Sudan zone in Nigeria, although dry-season minimum temperatures are lower. As in Nigeria, the land is covered by a deep drift of desert sand laid down in a dry Quaternary period; unlike Nigeria, however, this area has hitherto been sparsely populated and the vegetation still approximates to the virgin state. In 'mutemwa' vegetation there is a dense tangled understory of erect and scandent deciduous shrubs (e.g. *Dalbergia glandulosa, Combretum* spp.) which almost completely excludes grass. The upper-stratum trees are up to 20 m. high, and in places form a closed canopy, but usually are more widely spaced; typical species are *Baikiaea plurijuga* and *Pterocarpus stevensonii.* If fire enters, the underwood burns fiercely and is subsequently replaced by grass, and the 'mutemwa' gradually changes to open savanna very like that of the Sudan zone in Nigeria.

All this suggests that the climax vegetation of the Zurmi area consisted of a dense understory of deciduous dry-zone shrubs and lower-stratum trees with upper-stratum trees forming a rather open canopy. Although in itself stable, such vegetation would obviously be quickly destroyed once fires started, and would become degraded to an open savanna type with grass dominant. This is happening in Sesheke today, but may well have happened in Nigeria long ago. Further research may, however, indicate that the Zurmi area has been inhabited throughout the present climatic era, and that the vegetation has never been allowed to attain its climatic climax over continuous areas. Certainly today, ecological studies in the Sudan zone of Nigeria must consist largely of unravelling the varied effects of many kinds of biotic disturbance.

REFERENCES

[1] CLAPPERTON, H. and DENHAM, D. (1828) *Travels and Discoveries in Northern and Central Africa, 1822–24* (London).
[2] BARTH, H. (1857) *Travels and Discoveries in North and Central Africa* (London).
[3] HOGBEN, S. J. (1930) *The Muhammedan Emirates of Nigeria* (Oxford).
[4] NIVEN, C. R. (1937) *A Short History of Nigeria* (London).
[5] MORISON, C. G. T., HOYLE, A. C., and HOPE-SIMPSON, J. F. (1948) 'Tropical soil-vegetation catenas and mosaics', *J. Ecol.*, XXXVI.
[6] BRAUN, L. (1935) 'The undifferentiated deciduous forest climax and the association-segregate', *Ecology*, XVI.
[7] CHEVALIER, A. (1933) 'Le territoire géo-botanique de l'Afrique tropicale nord-occidentale et ses subdivisions', *Bull. Soc. bot. Fr.*, LXXX.

5 The Montane Vegetation of New Guinea

R. G. ROBBINS

From 'The montane vegetation of New Guinea', *Tuatara*, VIII 3 (1961) 124–33.

THE LOWER MONTANE FORESTS

ALMOST invariably a climb of a thousand feet or more has to be made to pass the limits of cultivation and enter the forest edge at the 6000 to 8000-ft level. Usually the lower zone is dominated by several evergreen oaks bearing simple mesophyll leaves and large acorns. *Castanopsis acuminatissima* is the most frequent species and *Lithocarpus moluccana* also is common. Above the oak zone, which is now largely cleared for gardens, the forest is either a *Nothofagus* beech forest (Plate 5) or a mixed podocarp–broadleaf forest. Lower montane rain forest in New Guinea may begin at altitudes as low as 3000 ft, but there is wide variation in its composition throughout the island and the formation still requires much study. It is in such forest of the inland mountain regions, however, that one finds so many close relatives of species which are common in New Zealand rain forests; the cool mid-mountain slopes in New Guinea have afforded a refugeum over past climatic fluctuations, and clearly there has been much two-way migration between the two areas. The palms, lianes and buttress roots of the New Guinea lowland forest are much reduced, while tree ferns, filmy ferns and bryophytes abound. Here on a fallen log the shiny red berries of a rampant *Nertera* may catch the eye. *Elatostema*, *Dawsonia* and familiar ferns may cover the forest floor, while among the shrubs species of *Piper*, *Pittosporum*, *Schefflera* and *Geniostoma* are immediately identified. Even small trees belonging to *Quintinia*, *Rapanea* (*Suttonia*) and Winteraceae (*Drimys* and *Bubbia*) are unmistakable, while in the canopy are Podocarpaceae, Elaeocarpaceae, Cunoniaceae, Lauraceae and Myrtaceae, all broadleaf families which dominate in New Zealand's mixed forests. The New Guinea montane forest, however, is far richer in genera and species the majority of which would be strange to a New Zealand botanist. The presence among these of such genera as *Carpodetus*, *Litsea*, *Melicope*, *Olearia*, *Ascarina*, *Paratrophis* and others give, as it were, but a New Zealand flavour to the vegetation. The gymnosperm element seldom reaches more than a stocking of eight to ten trees per acre, although they constitute a valuable and sought-after softwood timber resource.

Podocarp species which are common in the lower montane forest are *Podocarpus ledermannii*, *P. rumphii*, *P. imbricatus*, *P. amarus* and *P. neriifolius*. The last three of these are the most frequent and seedling regeneration is active. *P. amarus* and *P. neriifolius* are both types with broad flat leaves and may be found occasionally in the lowland forests. Two species of *Dacrydium* also occur in the highlands. *Dacrydium falciforme* is a small tree resembling the New Zealand miro, while *D. elatum* is close to rimu in appearance. Both, however, are local to rare in the montane forests. *Phyllocladus hypophyllus* and *Papuacedrus* (until recently *Libocedrus* sp.) belong more to the high montane forests.

The canopy of the lower montane forest is made up of a number of broadleaf species often reaching 100 ft or more in height. Genera such as *Opocunonia*, *Schizomeria* and *Aistopetalum* belonging to the Cunoniaceae are frequent. Elaeocarpaceae contribute many species of *Elaeocarpus* and *Sloanea*, while several *Cryptocarya* spp., a genus close to *Beilschmiedia*, represents the Lauraceae. Also here are *Syzygium* spp. (cf. *Eugenia*), and the lesser associates, *Albizia*, *Astronia*, *Alphitonia*, *Elmerrillia*, *Fagraea*, *Ficus*, *Guioa*, *Garcinia*, *Galbulimina*, *Ilex*, *Planchonella*, *Prunus*, *Perottetia*, *Sterculia* and *Zanthoxylum*.

In the second tree stratum with limits between 40–60 ft are to be found *Elaeocarpus*, *Ackama*, *Weinmannia*, *Pullea* and *Gillbeea*, all of the Cunoniaceae, *Ficus* and *Syzygium* spp., and a host of small trees such as *Sphenostemon papuanum*, *Sloanea*, *Sericolea*, *Myristica*, *Daphniphyllum*, *Dillenia montana*, *D. schlechteri*, *Diospyros*, *Quintinia*, *Casearia pachyphylla*, *Litsea*, *Couthovia*, *Zanthoxylum*, *Discocalyx*, *Rapanea*, *Ardisia*, *Pittosporum pullifolium*, *Helicia microcarpa*, *Timonius*, *Evodia*, *Eurya*, *Gordonia papuana*, *Ternstroemia* and the giant stinging tree, *Laportea*.

A dense shrubby layer includes the following: *Olearia*, *Casearia angiense*, *Bubbia*, *Aglaia*, *Dichroa febrifuga*, *Rhododendron*, *Geniostoma*, *Mearnsia cordata*, *Decaspermum* and *Xanthomyrtus*.

Members of the Rubiaceae such as *Amaracarpus*, *Psychotria*, *Mussaenda* and *Gardenia* are very frequent, while Melastomataceae is represented by *Medinilla*, *Everettia* and *Poikilogyne*. Small *Prunus* spp., *Eurya meizophylla* and the ubiquitous *Symplocos*, also *Pittosporum berberidoides*, *P. inopinatum*, a new species discovered during the survey, *Acronychia*, *Rhamnus*, *Evodia*, *Perottetia moluccana*. *Chloranthus* and *Ascarina* are among the sub-shrubs.

Small palms, *Pandanus* spp., tree ferns such as *Cyathea contaminans*, *Dicksonia* and a tall *Marattia* as well as masses of climbing bamboo all contribute to the forest structure.

A rich ground flora includes *Begonia*, some *Araceae*, the forest sedges *Carex*, *Scleria* and *Schoenus*, the dwarf *Pittosporum sinuatum*, while several species of *Pilea* and *Elatostema* are particularly abundant. *Alpinia* is a common Zingiberaceae together with the two Rubiaceous herbs *Ophiorrhiza* and *Argostemma*. *Piper* species may be small shrubs or semi-climbers. On the floor and fallen logs are a *Pratia* sp., and *Nertera granadensis*, while *Dianella* is frequent. Among the many ground ferns are to be found species belonging to the genera *Asplenium*, *Blechnum*, *Adiantum*, *Hymenophyllum* and *Leptopteris*. Here too is an abundance of lycopods and bryophytes.

Among the small climbers and scramblers present are *Parsonsia*, two species of *Freycinetia*, *Rubus*, *Clematis*, *Smilax*, *Celastrus*, *Alyxia*, *Hoya* and *Secamone* as well as members of the Gesneriaceae, Monimiaceae and Bignoniaceae.

Epiphytes include *Pittosporum ramiflorum*, *Schefflera*, climbing ferns and orchids (including *Dendrobium* and *Bulbophyllum*) and parasitic Loranthaceae.

BEECH FOREST

An excellent account of the discovery of extensive *Nothofagus* forests in New Guinea will be found in van Steenis [2]. Some sixteen species are now recognised, although the question of hybrids remains. With further recent collections, including those made by the C.S.I.R.O., a revision is already due.

Most of the New Guinea beeches are relatively large-leaved and with entire margins. However, on fruiting morphology van Steenis traces affinities with the New Zealand *N. fusca* group.

Beech forest has been recorded from about 3000 ft to 10,000 ft in New Guinea, but in the highlands its range was found to be between 7500 ft to 9000 ft (Plate 5). It descends, however, as a mixed element down to 4500 ft in the eastern highlands. *Nothofagus* was not found to play any part in the formations above 9000 ft and its true status appears to be as a member of the lower montane rain-forest formations.

The New Guinea beech forest is a two-layered forest with a closed canopy of spreading and interlocking crowns which may consist of 75 per cent *Nothofagus* species 90–100 ft high. It often forms a mosaic with the mixed broadleaf–podocarp forest occupying only one flank of the range or dominating on the ridges. In July the reddish flush of young leaves distinguished this pattern from some distance away, while the distinctive crowns show up well on aerial photos flown at 25,000 ft. In the highlands the largest continuous areas of beech are to be found along the southern Kubor Range and it is virtually absent from the ranges flanking the northern periphery of the valley. The beech trees are seldom found as scattered individuals but even in mixed forest form small isolated gregarious stands. No doubt some interesting correlation with migration paths and environment as is evident in New Zealand will emerge when the full pattern of the New Guinea *Nothofagus* forest is mapped.

In these New Guinea mountains, beech forests exist side by side, as in New Zealand, with a mixed broadleaf–podocarp forest with usually a sharp boundary between the two. However, here one is in a montane forest of the tropics and the aspect is of a more luxurious and humid forest than in a subantarctic beech forest in New Zealand. There is less contrast with the adjoining mixed forest and indeed many of the subordinate species overlap. The fagaceous trees dominate and have a similar layered appearance and even-formed canopy as in New Zealand, but mixed under-layers including *Pandanus* and climbing forest bamboo are present. The leaf litter is not heavy and topsoil, while tending to be acidic, is closely approximate to that in the mixed forest. Ferns, herbaceous plants and bryophytes may form a rich ground cover. The masses of climbing bamboo and the palm-like, stilt-rooted *Pandanus*, present also in the mixed forest, impart a distinctive physiognomy to the beech forests of the New Guinea highlands.

Species to be found in the upper stratum include several *Nothofagus* species, *Ackama*, and other Cunoniaceae, *Quintinia*, and *Podocarpus* with oaks, *Cryptocarya*, *Zanthoxylum* and *Alphitonia*.

The second stratum at about 30 to 50 ft may be sparse although rich in species. The smaller trees here include *Nothofagus*, *Mischocarpus*, *Weinmannia* and *Pullea*. Also several *Ficus* species, the Rubiaceous *Timonius* and *Psychotria*, two species of *Drimys*, *Rapanea* and other *Myrsinaceae*, *Syzygium* spp., *Evodia* (cf. *Melicope*) and *Couthovia*, a *Schefflera*, and *Sloanea* (Elaeocarpaceae).

Common members of the shrub layer are *Symplocos*, *Sphenostemon papuanum*, *Polyosma*, *Medinilla*, *Piper*, *Pittosporum pullifolium*, *Helicia microcarpa*.

Epiphytes are quite frequent, consisting mainly of pteridophytes, a small *Freycinetia*, orchids and scrambling Gesneriaceae.

The ground layers include ferns such as *Blechnum* and *Hymenophyllum*, with *Pilea* and *Elatostema*, terrestrial orchids, lycopods and bryophytes.

MONTANE FOREST

At 9000 ft above sea-level there is a change, often quite abrupt, in the forest formation. Above this level in the highlands are the regions of prolonged daily cloud cover and the so-called 'mossy' forest commences.

A single tree-layered community, montane forest is dominated by species of Myrtaceae and Podocarpaceae. The canopy is compact and low, being 35–40 ft high, and made up of close slender trees often crooked in growth. The branches and indeed the whole forest floor are typically festooned and carpeted with masses of bryophytes, both mosses and liverworts.

A bamboo and screwpines of the lower montane zone forest are now absent altogether and filmy ferns and tree ferns find true expression. All tropical features have gone and the atmosphere is damp and dripping. The leaves are smaller, coriaceous and dark green, giving a drab appearance to the forest.

Myrtaceae are well represented by species of *Decaspermum*, *Xanthomyrtus*, *Eugenia* and *Syzygium*. Also *Schefflera*, *Daphniphyllum*, *Quintinea*, *Elaeocarpus azaleifolius*, *Rapanea*, *Carpodetus*, *Drimys*, *Schuurmansia*, *Eurya*, *Prunus* and Rutaceae.

The mountain podocarps include *Podocarpus compactus*, *P. brassii* and *P. pilgeri*. *Papuacedrus* and *Phyllocladus* are also frequent.

A sparse shrub layer may include *Polyosma*, *Rhododendron* spp., *Vaccinium* spp., *Amaracarpus*, *Symplocos*, *Pittosporum* spp., *Piper* and appearing here for the first time, a low *Coprosma*.

Orchids, ferns and bryophytes make up the ground layer, which, in more open glades may include *Acaena anserinifolia*, *Libertia pulchella*, *Uncinia* sp., *Elatostema* and *Pterostylis*.

SUBALPINE SCRUB

At altitudes of 11,000 ft the upper limits of the montane forest are fringed with a dense, low, woody vegetation or subalpine scrub, patches of which may extend out into the alpine grassland.

The closed shrub layer, 15–20 ft high, consists of many *Rhododendron* and *Vaccinium* species together with *Acronychia*, *Olearia*, *Saurauia*, *Schefflera*, *Sericolea*, *Symplocos*, *Prunus*, *Decaspermum*, *Rapanea*, *Quintinea*, *Coprosma*, *Psychotria*, and *Amaracarpus*, *Pittosporum*, *Polyosma*, *Eurya* and *Drimys*.

These levels are above the prolonged mists of the montane forest zone and emergent *Podocarpus* and *Libocedrus* may occur, but there is little development of the high mountain forest formation described by Brass [1] around Lake Habbema in Dutch New Guinea and also reported for the Owen Stanley Range.

Mosses and liverworts are abundant in the subalpine scrub. Epiphytes include Loranthaceae, Orchidaceae and Pteridophyta, while as a ground cover are *Polystichum*, *Belvisia*, *Blechnum* and the herbs *Trigonotis*, *Libertia pulchella*, *Cardamine altigena*, *Acaena anserinifolia*, *Triplostegia glandulifera*, *Oxalis magellanica*, *Rubus moluccanus*, *Lycopodium*, *Galium*, *Epilobium* and *Geranium*.

ALPINE GRASSLAND

Above the tree-line at 11,000 to 12,000 ft the vegetation becomes tussock grassland. Extensive alpine grasslands occur on the summit areas of Mt Giluwe, Mt Hagen (Plate 6) and Mt Wilhelm [3] and make an appearance on some of the higher isolated peaks of the two ranges. Here a New Zealand botanist is quite at home. Large tussocks of *Hierochloe longifolia* and *Deschampsia klossii* dominate, forming clumps 2–3 ft high. In between grow the smaller tuft grasses represented by *Danthonia archboldii*, *D. vestita*, *D. schneideri* and *D. semiannularis*.

Other grasses recorded are *Deyeuxia*, *Agrostis*, *Anthoxanthum*, *Festuca papuana* and the minute *Poa crassicaulis* and *P. callosa*.

Alpine herbs, many of which form cushion plants, grow among the tussocks in a mat of mosses and lichens. Here may be found *Centrolepis philippinensis*, *Hydrocotyle sibthorpioides*, *Oreomyrrhis papuana* (syn. *andicola*), two species of *Potentilla* and several minute blue-flowered *Gentiana*. Of interest is the terrestrial *Astelia papuana* which is possibly identical to the New Zealand *A. linearis* of similar habitat.

Several species of *Epilobium*, a *Pratia* and *Euphrasia rectiflora* are present, while mountain daisies are represented by the genera *Anaphalis, Gnaphalium, Tetramolopium* and *Keysseria*.

Ferns include *Gleichenia*, the endemic *Papuapteris linearis* with *Schizaea fistulosa* as a new record. *Lycopodium scariosum* and the sedges *Carex philippinensis* and *Schoenus maschalinus* find representation, while groups of the small cycad-like *Cyathea tomentossissima* are characteristic of the depressions. Over much of the valley floors (many on Mt Wilhelm are glacial) the soil is peaty and poorly drained, resulting in alpine bog in which may be found *Drosera, Ranunculus, Potentilla, Astelia, Trachymene, Oreomyrrhis, Viola, Epilobium, Sagina* and *Carpha alpina* together with small *Gahnia, Carex, Agrostis* and *Danthonia*. Here also are *Scirpus aucklandicus* and *S. maschalinus*.

On higher slopes the soil is drier and more shallow with rocky outcrops. The tussocks thin out and small woody shrubs include *Styphelia, Gaultheria mundula, Drimys brassii, Detzneria tubata, Kelleria* (syn. *Drapetes*) *papuana, Coprosma, Eurya brassii* var. *erecta, Hypericum macgregorii, Hebe albiflora* and *H. celiata*. The last-mentioned are not whipcord hebes but similar to the New Zealand *H. catarractae*.

At the highest point on Mt Wilhelm, 14,950 ft above sea-level, flowering plants still form a sparse rock crevice community between the moss-covered granodiorite rocks. Here the mountain mosses *Andreaea rupestris* and *Rhacomitrium lanuginosum* var. *pruinosum*, which occur together with *Rhacomitrium crispulum*, were collected for the first time in New Guinea. All are common New Zealand alpine mosses. The plants recorded here are the grasses *Danthonia vestita, Poa callosa, Festuca papuana, Deyeuxia, Agrostis;* the ferns *Papuapteris linearis, Gleichenia bolanica* and the small herbs, *Cerastium keysseri, Lactuca, Ischnea elachoglossa* and *Trigonotis papuana*. While again there are sufficient New Zealand genera and even species to 'flavour' the vegetation strongly, an analysis of the Compositae and the Ericaceae, for example, shows a marked difference and reduces the apparent affinity.

CONCLUSION

The vegetation of montane New Guinea is only just becoming known. Already some confusion in the available literature reflects the great variation in local sites throughout this great island – hence the present account must be taken as referring essentially to the central highlands in Australian New Guinea.

The fascination of these montane forests lies in their affording a key to past southern floras. Thus the mixed lower montane forests of the highlands may be regarded as tropical counterparts of New Zealand's podocarp–broadleaf forests of tawa–hinau–kamahi and scattered rimu–totara–matai and miro.

In addition it contains many genera and species lost in New Zealand during the general post-Miocene regression, as, for example, the warmth-loving *Nothofagus brassii* group.

In the central highlands the major plant formations encountered are:

Lower montane rain forest	3000 to 9000 ft
Montane forest	9000 to 11,000 ft
Subalpine scrub	11,000 to 12,000 ft
Alpine grasslands	11,000 to 14,000 ft

REFERENCES

[1] BRASS, L. J. (1941) 'The 1938–39 expedition to the Snow Mountains, Netherlands New Guinea', *Journ. of the Arnold Arboretum*, XXII 271–342.
[2] VAN STEENIS, C. G. G. J. (1953) 'Papuan Nothofagus', *Journ. of the Arnold Arboretum*, XXXIV 4, 301–74.
[3] HOOGLAND, R. D. (1958) 'The alpine flora of Mount Wilhelm (New Guinea)', *Blumea*, suppl. IV, Dr H. J. Lam Jubilee Volume, pp. 220–38.
[4] ROBBINS, R. G. (in press) 'Montane formations in New Guinea', *UNESCO Symp. on Veg. of Humid Tropics, Bogor, Indonesia, 1958* (in press).

6 The Vegetation of the Imatong Mountains, Sudan

J. K. JACKSON

From 'The vegetation of the Imatong Mountains, Sudan', *J. Ecol.*, XLIV (1956) 341–3, 361–73.

THE Imatong mountains (Fig. 6.1) lie on the Sudan–Uganda border between 3° 40′ and 4° 20′ N. latitude, and 32° 30′ and 33° 10′ E. longitude, rising to a height of 3187 m. in the peak of Mt Kinyeti. The vegetation of the higher parts is a northward extension of that of the East African mountains, and it was felt that a detailed description of the vegetation of the Imatongs would be of interest for comparison with that of other areas.

The first description of the vegetation of the mountains was published by Chipp [1]; other short accounts have been given by Andrews [2] and Smith [3]. Vidal-Hall [4] discussed the silviculture of the forests, particularly the effect of fire on them. Dawkins [5] described the tropical forest at the foot of the mountains.

TOPOGRAPHY

The highest part of the mountain mass lies in the south-east, and consists of high ground with a group of peaks rising to 3000 m., more or less, the highest being Mt Kinyeti, 3187 m. From this central boss ranges run north-west, west and south-west, the first two being separated by the valley of the Kinyeti river, and the other two by the valley of the Ateppi. These ranges have a general level of rather over 2000 m., with peaks rising to 2400 m. From the fairly level tops of these ranges the ground drops steeply to the surrounding plains, which rise from an altitude of 600 m. at Torit to rather over 1000 m. on the Sudan–Uganda border. The plains are by no means flat or level, and are crossed by numerous streams separated by rounded ridges; small gneiss hills also frequently occur as outliers from the main mountain mass.

These three main levels, 3000 m., 2000–2500 m., and that of the present plain may correspond to different stages in the erosion of an ancient peneplain.

The configuration of the ground causes streams to tend to flow

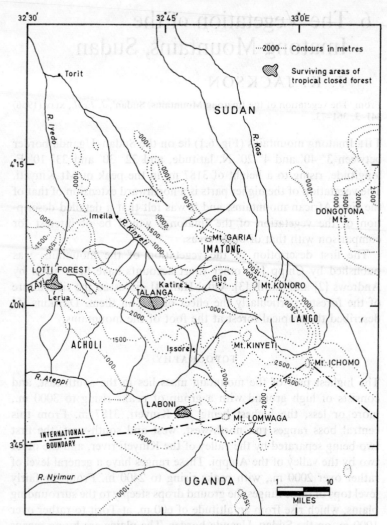

*Fig. 6.1   Map of the Imatong Mountains, adapted from Sudan Surveys
Department 1:250,000 maps*

gently in their upper reaches and they sometimes form swamps.
Below 1800 m., however, they drop rapidly in a series of waterfalls
to the plains. The transition from mountain-slope to plains is often
abrupt apart from a gently sloping detritus apron through which, in
places, the streams have cut deep ravines [5].

The streams rising in the Imatongs provide water for a large area in the south Sudan and northern Uganda, and the protection of their catchments is of considerable importance.

RAINFALL

The plains to the east and north-east of the mountains are rather dry: no figures are available from this area, but from the nature of the vegetation the rainfall would appear to be rather less than that of Torit, or say about 800 mm. per annum. As the slopes of the mountains are ascended the rainfall increases, until Gilo, at an altitude of 1900 m., has a rainfall of 2261 mm., the highest recorded in the Sudan. The Kinyeti valley and the western slopes of the mountains also have a rather high rainfall, averaging about 1500 mm., and then the rainfall decreases again in the western plains to about 1000 mm. at 25 km. from the foot of the mountains.

The rains generally break in March or early April, though there may be occasional showers, sometimes fairly heavy, in January and February. May is usually wet, but in June there is frequently a break in the rains of two or three weeks, which corresponds to the interval between the 'great' and 'small' rains of regions further south. After this break the rains continue until September or October, and in some years well into November or even early December. It frequently happens that several days of heavy rain occur in November or December after a spell of some weeks of dry weather.

The length and severity of the dry season vary considerably from year to year and have important ecological consequences, particularly in their effect on the incidence of fires. In some years there may be only two really dry months, while in others the drought period may extend to nearly five months. Thus in 1942–3 only 6·1 mm. fell at Katire between 4 October and 24 February, and no rain at all was recorded at Gilo between December 1945 and 20 March 1946; in the latter period there were particularly bad fires.

Dew is abundant at all altitudes especially in the dry season, and its occurrence undoubtedly reduces the effect of the drought at this period.

Mists are very frequent at the higher altitudes (that is, above the base of the daily cumulus cloud) and their effect is to be seen in the dense growth of *Usnea* on trees and rocks, and in the abundance of epiphytic ferns and mosses.

THE LOWER MONTANE FOREST ZONE, 1800–2600M.

This zone is characterised by the occurrence of forest containing *Podocarpus milanjianus* mixed with a number of other species, the most important of which are *Olea hochstetteri* and *Syzygium* sp. aff. *S. gerrardii*.

This climax forest does not, however, occupy the whole of the zone. The summits of peaks and ridges are frequently of bare rock, while below them lies grassland. Below 2300 m. this grassland has scattered trees of *Protea gaguedi*, while above this altitude it contains few trees as long as it is burnt annually. Forest often extends to within 20 or 30 m. vertical distance of the hill crests, and in this case there is often a narrow transitional zone between the forest and the grassland in which *Hagenia abyssinica* is characteristic. When the grassland is protected from fire the *Hagenia* spreads into it rapidly, followed by forest species.

In other areas, particularly at lower altitudes, there is a more gradual transition from grassland through *Acacia abyssinica* woodland and *Albizia gummifera* woodland to *Podocarpus–Syzygium* forest.

Along streams species of riparian character occur and there are small areas of swamps. Also, particularly along streams and in valleys, there are fairly large areas of broken forest, which appear to be post-climax, occurring after the trees of the climax forest have fallen.

A more detailed description of the different seral stages follows.

Rocky areas with Protea and Loudetia

The first colonisers of rocks are lichens, followed by mosses and *Selaginella njamnjamensis*; in the mat of vegetation produced by these species *Eriospora schweinfurthiana*, *Sporobolus centrifugus*, *Aeolanthus heliotropioides* and *Eragrostis hispida* establish themselves. In cracks in the rock *Coleus lactiflorus*, *Aloe* sp. and occasional *Hymenodictyon floribundum* occur.

The next stage is grassland of *Loudetia arundinacea* var. *trichantha* among which are scattered fire-gnarled trees of *Protea*. When this grassland is burnt annually, as was the case until recent years, its composition remains fairly constant, being at the end of the rainy season a sea of *Loudetia* 1 to 1·5 m. high with fire-blackened *Protea* among it. Below 2000 m. the *Protea gaguedi* is accompanied

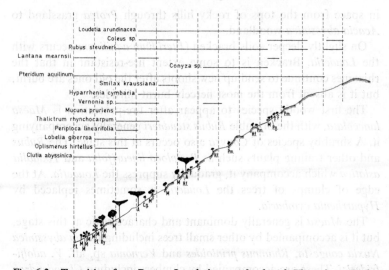

Fig. 6.2 Transition from Protea–Loudetia *grassland with* Hagenia *to* Podocarpus–Syzygium *forest. Site: Tipo Hill. Altitude of summit 2,300 m. approx. Scale (horizontal and vertical) 1:2,500. (The broken vertical lines are at a distance apart representing 20 m.) Record made April 1954. Area last burnt in 1946.*

G, Grumilea *sp.; H*, Hagenia abyssinica*; M*, Maesa lanceolata*; N*, Nuxia congesta*; Oc*, Ochna holstii*; O.h.*, Olea hochstetteri*; P*, Podocarpus milanjianus*; P.g.*, Protea gaguedi*; S*, Syzygium *sp. aff. S. gerrardii.*

The occurrence of shrubs and herbs is represented by the horizontal lines at the top of the diagram.

by *P. madiensis*, and at this altitude it is accompanied by *Faurea speciosa* and two species from the plains savannah, *Combretum gueinzii* and *Erythrina abyssinica.*

Two seasonal aspects can be distinguished. After the grass is burnt a number of geophytes come into flower, as in the plains. These include *Gladiolus quartinianus, Hypoxis multiflorus, Wahlenbergia silenoides, Delphinium leroyei, Lightfootia abyssinica* and a number of orchids. In the other seasonal aspect when the grass is tall a number of herbs of about the same height as the grass come into flower; these include *Echinops amplexicaulis, Guizotia scabra* and *Lantana mearnsii.*

The *Loudetia* is locally replaced by *Setaria sphacelata*, especially where the soil is somewhat deeper.

When this grassland is protected from fire a succession towards *Acacia abyssinica* woodland begins. A similar succession is also seen

in space from the tops of rocky hills through *Protea* grassland to
Acacia abyssinica woodland.

On slightly deeper soils bracken (*Pteridium aquilinum* occurs with
the *Loudetia*. Bracken is to some extent fire-resistant in that the
rhizomes continue to send up new shoots after the old ones are burnt,
but it is absent from the most fiercely burnt hillsides.

The first woody species to appear after fire-protection is *Maesa
lanceolata*, with the bramble *Rubus steudneri* generally accompanying
it. A shrubby species of *Conyza* also occurs at this stage. The *Rubus*
and other trailing plants such as *Periploca linearifolia* and *Toddalia
asiatica* which accompany it, gradually suppress the *Loudetia*. At the
edge of clumps of trees the *Loudetia* is sometimes replaced by
Hyparrhenia cymbaria.

The *Maesa* is generally dominant and characteristic at this stage,
but it is accompanied by other small trees including *Clutia abyssinica
Nuxia congesta*, *Rhamnus prinioides* and *Vernonia* sp. aff. *V. adolfi-
friderici*. These are accompanied by climbers including *Clerodendron
johnstonii*, *Smilax kraussiana*, *Gouania longispicata* and *Phytolacca
dodecandra*. The giant lobelia, *Lobelia giberroa*, an erect woody herb
up to 3 m. high, is common in these situations. Other herbs include
Alectra senegalensis, *Leonotis velutina*, *Plectranthus defoliatus*,
Cycnium tomentosum, *Micromeria punctata* var. *purtschelleri*,
Helichrysum hochstetteri, *H. schimperi*, *Guizotia scabra*, *Bidens
imatongensis*, *Thalictrum rhynchocarpum*, *Clerodendron myricoides*
var. *discolor*, *Pycnostachys meyeri* and *Osbeckia abyssinica*. Grasses
are sparse, the principal species being *Panicum* sp. aff. *P. monticolum
Poa leptoclada* and *Brachypodium* sp. aff. *B. simplex*.

Young trees of *Acacia abyssinica* now begin to appear, beginning
the next stage in the succession.

Above about 2300 m. *Protea gaguedi* becomes rare, though it is
found occasionally up to 2500 m. At these higher altitudes, and occasionally down to about 2000 m., *Hagenia abyssinica* becomes the
most important colonist of grassland, especially where forest is
separated from grassland by a narrow transitional zone. The
Hagenia occurs with the *Maesa* and at higher altitudes replaces it;
at these higher altitudes *Tephrosia atroviolacea* is also an important
colonist. *Hagenia* may also be succeeded by *Acacia abyssinica*,
but generally passes into *Podocarpus–Syzygium* forest direct, without
the *Acacia abyssinica* stage.

Acacia abyssinica *woodland*

This type of woodland which follows *Maesa* and *Rubus* in the succession from grassland towards forest is generally found between 1500 and 2100 m., but on the north-east side of the mountains occurs between 1700 and 2500 m. In some parts of the mountains it does not occur and here the succession appears to be direct from *Protea* grassland to *Albizia gummifera* woodland or *Podocarpus* forest. *Acacia abyssinica* woodland is also especially characteristic of areas which have been cultivated in the past.

In typical *Acacia abyssinica* woodland the very flat-topped Acacias have beneath them a dense tangly herbaceous undergrowth, the species in the main being those following *Protea gaguedi* in the succession. The following is a typical list, taken at 1900 m. altitude:

Trees
　Acacia abyssinica
　Croton macrostachys
　Maesa lanceolata
　Nuxia congesta
　Albizia gummifera ⎫
　Albizia maranguensis ⎬ Young trees from next stage in the
　Polyscias fulva ⎭ succession

Shrubs and woody herbs
　Lobelia giberroa
　Lantana mearnsii
　Vernonia sp.
　Emilia sagittata
　Triumfetta sp.

Climbers
　Cissampelos rigidifolia
　Periploca linearifolia
　Clerodendron johnstonii
　Toddalia asiatica
　Mucuna pruriens
　Rubus steudneri

Herbs
　Pteridium aquilinum
　Thalictrum rhynchocarpum
　Pentas carnea

Acacia abyssinica is to some extent fire-resistant when mature and it often happens that a severe fire has partially reversed the succession and left the *Acacia abyssinica*, without its shrub layer, over open grassland.

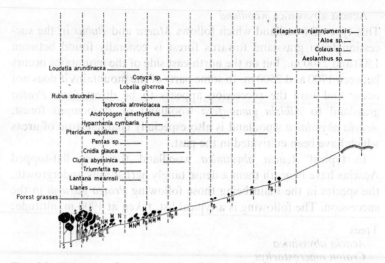

Fig. 6.3 Transition from Protea–Loudetia *grassland to* Acacia abyssinica *woodland. Site: Observation Hill, Gilo. Altitude of summit 2,000 m. Scale (horizontal and vertical) 1:2,500. (The vertical dotted lines are at the equivalent of 20 m. apart.) Record made in April 1954. The area was last burnt in autumn 1945.*

Symbols for trees: A, Acacia abyssinica; *C.g.,* Combretum gueinzii; *F,* Faurea speciosa; *H.f.,* Hymenodictyon floribundum; *M.* Maesa lanceolata; *P.g.,* Protea gaguedi; *P.f.,* Polyscias fulva; *P.m.,* Protea madiensis; *Ry,* Rytiginia *sp. Rock outcrops are shaded.*

Albizia *woodland*

This type of woodland, characterised by the dominance of *Albizia gummifera* and, below 2000 m., of *A. maranguensis*, is between 1800 and 2100 m. generally a relatively short-lived stage in the succession from *Acacia abyssinica* woodland to *Podocarpus* forest. (It occasionally succeeds *Protea–Maesa* direct without the intervening *Acacia abyssinica* stage.) Below the lower limit of *Podocarpus*, however, between 1600 and 1800 m., it occupies quite large areas in which the trees grow to great dimensions up to 25 or 30 m. in height by 80 cm. in diameter. At such altitudes it may possibly represent the climax.

Associated with *Albizia gummifera* are *Polyscias fulva, Croton macrostachys, Macaranga kilimandsharica,* occasional *Dombeya goetzenii,* and locally *Bridelia brideliifolia.* In some places the *Bridelia* is common and both it and the *Macaranga* may locally replace the *Albizia* as dominants. *Clausena anisata* is a common understorey tree. Relict trees of *Maesa lanceolata* and *Acacia abyssinica*, often moribund, may remain.

The tangle of herbs and scandent shrubs associated with *Acacia abyssinica* woodland is gradually weakened by the shade, until the ground flora becomes scanty, consisting chiefly of forest grasses such as *Oplismenus hirtellus*, *Brachypodium flexum* and *Pseudobromus sylvaticus*. At this stage colonisation by the species of the forest climax begins.

At lower levels more shrubs are associated with the Albizia woodland, and the following is an example recorded at 1700 m.:

> *Albizia maranguensis*
> *A. gummifera*
> *Dracaena steudneri*
> *Trichilia volkensii*
> *Clausena anisata*
> *Clerodendron johnstonii*
> *Fagara* sp.
> *Ochna* sp.
> *Monothecium* sp.
> *Brillantaisia* sp.

Albizia woodland is not usually susceptible to fire but is occasionally burnt by late fires at the end of the dry season. Large areas of such woodland were burnt in 1946; in some places the trees were not severely damaged and a dense growth of *Impatiens elegantissima* and *Girardinia condensata* followed the fire.

Albizia gummifera also occurs as a stage in the recolonisation of areas of post-climax broken forest.

Forest climax with Podocarpus, Olea *and* Syzygium

Climax forest occurs from the edge of streams (provided that the ground is not too swampy) almost to the summit of the ridges, but is perhaps most abundant on the upper parts of the slopes towards the ridge-tops. The description of the mountain forests as 'ravine forests' (e.g. Andrews [2]) is misleading.

The dominant trees are generally 20 to 25 m. high, and the canopy is closed. There is a moderately well-developed but not continuous second storey of trees 5 to 8 m. high and a shrub layer 2 to 3 m. high in which *Acanthus eminens* is dominant almost throughout. The ground flora (except where gaps occur in the canopy) is generally sparse and sometimes almost completely absent.

The following is a generalised list of the species occurring:

Dominant trees
  Podocarpus milanjianus (f., l.c.)
  Olea hochstetteri (c.)
  Syzygium sp. aff. *S. gerrardii* (c.)
  Olea welwitschii (o.) (*Steganthus welwitschii* in
    Andrews [6] vol. II)
  Ocotea kenyensis (o.) (*Ocotea viridis* in Andrews [6] vol. I)
  Ochna holstii (o.)
  Macaranga kilimandsharica (o.)
  Pygeum africanum (o.)
  Dombeya goetzenii (o.)
  Croton macrostachys (o.)
  Fagaropsis angolensis (r.)
  Allophyllus abyssinicus (r.)
  Fagara macrophylla (o.)
  Chrysophyllum fulvum (r.)

Second-storey trees
  Grumilea sp. (c.)
  Teclea nobilis (f.)
  Cassipourea sp. aff. *C. abyssinica* (o.)
  Maytenus lancifolius (o.)
  Ilex mitis (o.)
  Strychnos mitis (r.)
  Bersama abyssinica (r.)
  Ritchiea pentaphylla (r.)

Shrubs
  Acanthus eminens (a.)
  Galiniera coffeeoides (o.)
  Heinsenia diervilloides (o.)
  Dracaena afromontana (r., l.c.)

Climbers
  Clerodendron johnstonii (l.c.)
  Landolphia ugandensis (o.)
  Hippocratea sp. nr. *H. indica* (o.)
  Dalbergia lactea (r.)

Grasses and herbs
  Oplismenus hirtellus (f.)
  Pseudobromus sylvaticus (o.)

Erharta abyssinica (o.)
Brachypodium flexum (o.)
Pilea tetraphylla (o.)
P. ceratomera (o.)
Hypoestes sp. (o.)
Desmodium sp. (o.)

A rough idea of the proportions of the various dominants can be obtained from the following results of a strip survey to estimate timber volumes in typical forest of this type, at 2000 m. Only trees over 90 cm. in girth were recorded.

Species	Trees per hectare
Syzygium sp. aff. *S. gerrardii*	46
Olea hochstetteri	44
Podocarpus milanjianus	18
Olea welwitschii	6
Ocotea kenyensis	6
Ochna holstii	4
Teclea nobilis	4
Grumilea sp.	4
Croton macrostachys	2
Fagaropsis angolensis	2
Dombeya goetzenii	1

As the altitude increases so does the proportion of *Podocarpus*, until in the upper part of this type of forest *Podocarpus* is almost pure except for a little *Olea hochstetteri*.

Podocarpus milanjianus itself often acts as a colonising species. Conditions needed for establishment of its seedlings are absence of fire and severe competition from grass and herbs, and moderate, but not too heavy, shade. Thus *Podocarpus* seedlings may occur in any place at a suitable altitude where there is light shade and a fairly clean forest floor, whether at a relatively early stage in the succession, for instance below *Maesa*; in the *Albizia gummifera* stage; or under mature *Podocarpus* forest. Perhaps the commonest site where seedlings are found is in the ecotone of *Maesa*, *Hagenia*, etc., between grassland and forest. If *Podocarpus* occurs early in the succession the intermediate stages may be passed over.

Near Gilo large numbers of young *Podocarpus* were cut for poles. Elsewhere, however, and particularly at altitudes of over 2200 m. all ages of *Podocarpus* from seedlings to mature trees are found in the same area. Thus *Podocarpus* does regenerate itself under its own shade, and the climax forest would appear to be self-perpetuating, or at least capable of enduring for several generations of its constituent trees.

Olea hochstetteri and *Syzygium* sp. make their first appearance later in the succession than the *Podocarpus*, and at altitudes of between 1800 and 2000 m. *Syzygium* forest should perhaps be regarded as the climax.

Broken forest post-climax

Considerable areas are occupied by this type of vegetation, particularly in valleys and on south-facing slopes. The ground is covered by a tangled mass of climbing herbs 2 or 3 m. deep, beneath which lie the fallen stems of the trees which previously grew there. The few trees remaining standing are covered almost to their crowns with creepers, producing the effect of thick green columns.

Among the creepers Acanthaceae predominate, including *Isoglossa* sp. nr. *I. ovata*, *Mimulopsis solmsii* and *Hypoestes* sp., together with *Triumfetta macrophylla*, *Clerodendron johnstonii*, *Urera* sp. and occasional *Rubus steudneri*. A few small trees occur among the creepers, especially *Galiniera coffeeoides* and *Abutilon longicuspe*.

Fig. 6.4　Section through Podocarpus–Syzygium *forest. Site: Tipo. Altitude 2,200 m. approx. Scale 1:800. Recorded April 1954.*

G, Grumilea *sp.;* M.k., Macaranga kilimandscharica; M.l., Maytenus lacifolius; N, Nuxia congesta; Oc, Ochna holstii; O.h., Olea hochstetteri; O.w., Olea welwitschii; P, Podocarpus milanjianus; S, Syzygium *sp. aff.* S. gerrardii; T, Teclea nobilis.

Note: *There is rather more* Olea welwitschii *in this section than is typical.*

This type of vegetation is eventually recolonised by trees of which the most characteristic species are *Dombeya goetzenii* and *Croton macrostachys*, followed by *Albizia gummifera*, after which the climax returns.

In some places, particularly in very deep, rather dark valleys, the post-climax vegetation has rather a different composition with *Dracaena steudneri* and *Ensete edule* as the most characteristic species; with them are found *Crassocephalum mannii* and *Alangium chinense*, with Acanthaceous climbers, etc., as before.

The cause of these areas of broken forest is not known exactly; in some places they are found where a late fire has entered the forest, but it seems unlikely that this can be the cause of all such areas. Perhaps they are due merely to a collective senescence and eventual wind-blow of the trees of the forest climax.

Riparian vegetation
Swampy stream beds in this zone have the following species:

> *Panicum glabrescens* (l.d.)
> *Pennisetum trachyphyllum* (f.)
> *Polygonum aculeatum* (l.c.)
> *Ranunculus multifidus* (f.)
> *Droguetia iners* (o.)

with at the edges often wide beds of *Impatiens elegantissima*. Trees characteristically found near streams are *Schleffera abyssinica* and *Aningeria adolfi-friderici*, the latter especially in deep valleys. The tree fern *Cyathea deckenii* also occurs in marshy ground near streams, usually under heavy shade.

THE HIGHER MONTANE FOREST ZONE, 2600–3000M

In this zone *Podocarpus milanjianus* continues to form the climax, though here it is almost pure or mixed with a little *Olea hochstetteri*. Seral stages with *Rapanea neurophylla*, *Hypericum lanceolatum* and *Hagenia abyssinica* cover a large part of the area, and the bamboo *Arundinaria alpina* is locally dominant. Lichens, particularly *Usnea* spp., become very abundant and much of the forest is wet or swampy. There are large areas of fire-climax grassland, and the forest is often reduced to strips and patches among the mountain meadow so that the landscape has the appearance of a park with small coverts and copses.

Higher montane grassland
This type of vegetation covers large areas, particularly round Kippia. It has shortish, generally not more than knee-high grasses, mixed

with very numerous and abundant herbs. It includes frequent boggy areas.

Bulbostylis atrosanguineus, a wiry-leaved, tussock-forming sedge, is dominant over wide areas. The most common grass is *Exotheca abyssinica*, with *Digitaria uniglumis* and *Setaria sphacelata* also common; other species of grasses are *Andropogon amethystinus*, *Koeleria cristata* var. *brevifolia*, *Festuca griidula*, *Agrostis producta* and *Helictotrichon elongatum*.

Among herbs *Bidens mossii* (syn. *Coreopsis tripartita*) is abundant and at certain times of year the grassland is yellow with its flowers. Other herbs are *Moraea diversifolia*, *Hypoxis urceolata*, *Delphinium leroyei*, *Athrixia rosmarinifolia*, *Justicia whytei*, *Hebenstreita dentata*, *Lactuca capensis*, *Polygala abyssinica*, *Sopubia ramosa*, *Micromeria biflora*, *Rhamphicarpa tenuisecta*, *Lotus corniculatus*, *Plectranthus schimperi*, *Geranium simense*, *Euphorbia depauperata*, *Alepidea* sp., *Peucedanum winkleri*, *Cyanotis barbata*, *Kyllinga* sp. aff. *K. triceps*, *Lysimachia* sp., *Trifolium ruepellianum* var. *preussii*, *T. simense* and *Ranunculus oreophytus*.

Swampy areas in the mountain meadow are often muck swamps with deep wet black soil. In these swamps occur *Carex steudneri* and *C. fischeri*, *Juncus dregeanus*, *Eriocaulon* sp., *Scirpus corymbosus* and *Utricularia tribracteata*. In slightly less wet places are found *Lathyrus hygrophilus*, *Hypericum peplidifolium* and *Carduus* sp., while at the edges of the swamps grow *Gunnera perpensa*, *Setaria strata*, in thick clumps, and *Agrostis* sp.

In other swampy areas *Impatiens hochstetteri*, *Alchemilla* sp. and *Ranunculus oreophytus* are found.

Transition from grassland to forest

As at lower altitudes, when burning is stopped *Hagenia abyssinica* rapidly colonises the grassland, with *Tephrosia atroviolacea*, *Hypericum lanceolatum* (*H. leucoptychoides* in Andrews [6] vol. 1), and *Gnidia glauca*. These are followed by *Dombeya goetzenii Rapanea neurophylla*, *Maytenus lancifolius* and *Agauria salicifolia*, before finally *Podocarpus milanjianus* comes in. This type of transitional forest occupies considerable areas at higher elevations in the Imatongs, and seems to be analogous to the *Hagenia–Hypericum* zone of the East African mountains [7]. However, in the Imatong mountains at least it is not the climax but is succeeded by *Podocarpus* forest.

The trees of this transitional forest average about 8 m. in height and are rather widely spaced. The ground beneath them is often moist, in which case *Cyperus derreilema* is often dominant in the ground flora. With it on moist sites are *Veronica abyssinica, Alchemilla* sp., *Impatiens hochstetteri* and *Viola eminii*, while on drier sites *Asparagus asiaticus* var. *scaberulus, Lobelia giberroa* and *Rubus steudneri* occur. Common grasses are *Bromus runssoroensis* and *Brachypodium simplex.*

Another form of transitional vegetation is the fringe of shrubs and woody herbs separating forest from annually burnt grassland; this belt forms a protective zone preventing fires (unless exceptionally fierce) from entering the forest. In this zone are found *Tephrosia atroviolacea, Kalanchoe petitiana, Cyathula schimperiana, Cineraria kilimandscharica, Plectranthus schimperi, Coleus* sp., *Leonitis velutina, Bothriocline schimperi, Vigna schimperi, Berkheya spekei, Spergula arvensis, Justicia betonica, Carduus kikuyuorum, Sparmannia ricinocarpa* and *Caucalis incognita. Hyparrhenia cymbaria* is a common grass at the forest edge.

Higher montane forest climax

The climax at this altitude consists of almost pure *Podocarpus milanjianus*, with occasional *Olea hochstetteri*. The trees are stunted and rarely over 10 m. in height, and covered in *Usnea* and mosses. The ground flora again is of a very wet type with *Cyperus derreilema, Carex cyrtosaccus, Ranunculus multifidus, Impatiens elegantissima, Hypoestes triflora* and *Alchemilla* sp. *Clematis simensis* is a common climber. The most common grasses are *Helictotrichon thomasii* and *Ehrharta abyssinica.*

The areas occupied by this type of forest are relatively small and seral stages with *Dombeya and Rapanea* mixed with the *Podocarpus* are much more common.

Arundinaria alpina *bamboo forest*

The alpine bamboo, *Arundinaria alpina*, occurs in a discontinuous belt at about 2700 m., especially to the east and north of Mt Kinyeti. It is often mixed with *Podocarpus* forest. As in the case of the *Oxytenanthera* bamboo at lower altitudes the factors governing the distribution of *Arundinaria*, and its place in the succession, are rather obscure.

ERICACEOUS ZONE, OVER 3000 M.

This zone is found at the summit of Mt Kinyeti, above the line of *Podocarpus* forest at 2900 to 3000 m. Some of its characteristic species occur rather lower on exposed rocky peaks, indicating that exposure rather than altitude alone may be the chief factor producing this type of vegetation in the Imatongs.

Certain life-forms are very characteristic of this zone such as ericoid shrubs; wiry, often aromatic herbs; and plants with white hoary leaves. All these characters seem to be adaptations to the intense insolation and high rates of transpiration found at high altitudes near the equator.

Woody plants in this zone rarely attain more than 2 m. in height. The dominant shrubs are *Erica arborea*[1] and *Myrica salicifolia*; others are *Anthospermum usambarense*, *Senecio* sp. aff. *S. denticulata*, *Alchemilla argyrophylla*, *Gnidia glauca*, *Hypericum lanceolatum*, *Crassula pentandra* var. *phyturus*, *Erlangea imatongensis*, *Blaeria breviflora* and *Smithia volkensii*. Herbs include *Micromeria punctata* var. *purtschelleri*, *Calamintha simensis*, *Asparagus asiaticus* var. *scaberulus*, *Crassula alba*, *Hebenstreita dentata*, *Helichrysum argyranthum*, *Carduus theodori*, *Coreopsis chippii* and *Silene burchellii* (*Silene chirensis*, in Andrews [6] vol. i). The dominant grasses are *Exotheca abyssinica* with *Tripogon snowdenii*, *Festuca rigidula*, *Andropogon thomasii* and *Agrostis sororia*.

In slight depressions where a little water can collect, the following are found: *Hypericum lanceolatum* (here a stunted undershrub), *Anthospermum usambarense*, *Rubus* sp., *Dipsacus pinnatifidus*, *Pimpinella oreophila*, *Cyperus* sp., *Anemone thomsoni*, *Phyllanthus* sp. and *Lithospermum officinale*.

The actual summit of Kinyeti is of rock slabs and boulders, heavily festooned with *Usnea*. In the crevices of the rocks, and among the boulders, the following are found: *Anthospermum usambarense*, *Myrica salicifolia* (here only 50 cm. high), *Lobelia dissecta*, *Bartsia trixago*, *Helichrysum formosissimum*, *Adenocarpus mannii*, *Erlangea* sp., *Trifolium multinerve*, *Cyanotis barbatus*, *Viola eminii*, *Minuartia filifolia*, *Pimpinella oreophila*, *Alepidea* sp., *Micromeria punctata* and *Helichrysum fruticosum*. Grasses include *Pentaschistis imatongensis* and *Tripogon snowdenii*.

[1] It is possible, as pointed out by Hedberg [7], that other Ericaceae have been included with this species.

In the past even this Ericaceous zone was sometimes burnt, and so the vegetation is possibly fire-climax rather than true climax. However, in other parts of East Africa the shrubby vegetation of *Erica*, etc., is considered as climax at these altitudes (e.g. Thomas [8]) and there is no evidence of colonisation by other types in the Imatongs.

THE AFFINITIES OF THE VEGETATION OF THE IMATONG MOUNTAINS

The lowland forests
The areas of lowland rain forest at the foot of the Imatongs strongly resemble the 'Mixed Forest' seral stage of Budongo forest, Uganda, 250 km. to the south, though the *Cynometra* climax of Budongo is absent from the Imatongs [9]. The absence of the climax found in Uganda may possibly be due to the relative immaturity of much of the Imatong forests, together with their small areas which may not give sufficient room for the climax to develop.

Between Budongo and the Imatongs lies the Zoka forest in East Madi, which again appears to resemble the Imatong forests closely. Apart from this area there is little closed forest today between Budongo and the Imatongs, but there are considerable areas of savannah woodland which appear to have been derived from rain forest at a comparatively recent date, and it appears quite probable that all these forest areas are remnants of a great belt of forest which joined up with the Ituri forest of the Congo.

In the Sudan the nearest areas of forest comparable to the Imatong lowlands forest lie on the Aloma plateau some 200 km. to the west, but there is more resemblance between Budongo and the Imatong forests than there is between the Imatong forest and those of the Aloma.

Considered in a more general way, all these areas have considerable floristic affinities with the equatorial forest of the Congo and West Africa, of which they should be regarded as outliers.

The montane vegetation
In the south Sudan, vegetation similar to that of the Imatongs is found on two groups of mountains, the Dongotona mountains, and Mt Lotuke in the Didinga mountains, which lie 30 km. and 100 km. to the east of the Imatongs respectively. The vegetation of the Dongotona mountains differs only in minor detail from that of the

Imatongs and the two areas can be considered together [10]. (The record of *Podocarpus gracilior* from the Dongotonas, in Andrews [2], appears to be based on a misidentification.) Mt Lotuke differs somewhat, however, in that the area of moist *Podocarpus milanjianus* forest found on it is surrounded by a drier belt with remains of forest of *Juniperus procera* and *Olea chrysophylla*, now largely destroyed by fire [11]. Such a division between drier *Juniperus* forest and moister *Podocarpus* forest is common further south, for instance on Mt Kenya, but *Juniperus* does not occur in the Sudan west of the Didinga mountains. North of Mt Lotuke *Podocarpus milanjianus* also occurs in the Didingas as a constituent of evergreen thickets in grassland, where it is mixed with *Olea chrysophylla*, *Euclea divinorum*, *Scutia myrtina*, etc.

From Mt Lotuke a series of peaks form a broken chain connecting this area with Mt Elgon and the East African highlands; these are Morungole, Moroto, Napak and Kadam [8]. The vegetation of Morungole has considerable affinities with that of the Imatongs [8], but *Podocarpus* forest is absent, the dominant species being *Dombeya goetzenii*. This is a secondary species in the Imatongs, particularly common in areas where the *Podocarpus* forest has been destroyed, and it seems possible, therefore, that on Morungole also *Podocarpus* forest would have been the climax vegetation had it not been destroyed.

Moroto and Napak have no *Podocarpus* but the drier type of forest with *Juniperus* and *Olea chrysophylla*, but *Podocarpus* forest occurs again on Kadam. Here it is associated with *Chrysophyllum* spp., *Syzygium guineense*, *Antiaris toxicaria*, *Teclea nobilis*, *Croton macrostachys* and *Conopharyngia holstii;* thus although it has resemblances to the forest of the lower zone in the Imatongs it is by no means identical in composition. (It is possible that the *Syzygium guineense* referred to here and the *Syzygium* sp. of the Imatongs are the same, as the nomenclature of this genus is very confused.)

The vegetation of Elgon, at least in part, has close resemblances with that of the Imatongs. Lugard [12] describes a lower zone of *Acacia abyssinica* between 1800 and 2200 m. in which *Albizia gummifera* also occurs. This corresponds to a seral stage at similar altitudes in the Imatongs, not the climax. Above this *Acacia abyssinica* zone on Elgon, *Podocarpus* forest is found from 2200 to 3000 m. and *Erica*, etc., above it.

The 'scattered tree grassland' of Edwards and Bogdan [13] occurring in Kenya between 900 and 1800 m. altitude is very similar in composition to the fire-climax vegetation found on the slopes of the Imatong at these altitudes. The forest in Kenya most resembling the Imatongs lower montane forest is the 'Temperature Rain (Camphor) Forest' described by Dale [14], but the characteristic species of the Kenya forests, *Ocotea usambarensis*, is absent from the Sudan, though to some extent it is replaced by the allied *O. kenyensis*. In Tanganyika and the Congo also *Podocarpus* in this type of forest tends to be associated with *Ocotea usambarensis* [15, 16]. On Muhavura in Uganda Snowden recorded, as dominant species, *Albizia gummifera*, *Dombeya goetzenii*, *Polyscias ferruginea* and *Trema guineensis*. These correspond to the *Albizia gummifera* seral stage in the Imatongs, which there leads to a *Podocarpus–Syzygium* climax [17].

The dominance of *Syzygium* sp. aff. *S. gerrardii* which is such a feature of the lower montane forests of the Imatongs does not appear to occur elsewhere in East Africa, though some, at least, of the records of *S. guineense* may refer to the same species. In the Imatongs the '*Syzygium* sp. aff. *S. gerrardii*' of the mountains and the '*S. guineense*' of the plains appear to be quite distinct entities, the former differing from the latter in its pale-grey flaky bark and smaller leaves, besides its ecological characteristics.

The upper montane forest zone of the Imatongs is represented in most of East Africa by the bamboo and *Hagenia–Hypericum* zones of Hedberg [7]. The bamboo zone in the Imatongs is not as distinct as on some of the East African mountains, the bamboo being generally mixed with other species, especially *Podocarpus*, and the zone also is discontinuous. The *Hagenia-Hypericum* zone is represented in the Imatongs, but as a seral stage leading to *Podocarpus* forest; whereas the only mountain group mentioned by Hedberg where *Podocarpus* forest extends to 3000 m. is Kilimanjaro. (Lugard, however, records it from up to 3000 m. on Elgon; see above.) It appears possible that the *Hagenia–Hypericum* zone of some of the other East African mountains may also be a seral stage, produced from the original vegetation by human interference.

The Ericaceous zone, of which only a small area occurs in the Imatongs, appears to resemble the lower parts of the Ericaceous zone of other East African mountains, though there are a number of endemic species in the Imatongs among the grasses and herbs.

It is interesting how often the dominant vegetation recorded from other mountains in East Africa corresponds to seral stages and not the climax in the Imatongs. This may be because the effect of human interference has been less intense or less prolonged in the Imatongs, or it may be due to climatic factors.

REFERENCES

[1] CHIPP, T. F. (1929) 'The Imatong Mountains, Sudan', *Kew Bull.*, p. 177.
[2] ANDREWS, F. W. (1949) 'The vegetation of the Sudan', in J. D. Tothill (ed.), *Agriculture in the Sudan* (London).
[3] SMITH, J. (1949) 'Distribution of tree species in the Sudan in relation to rainfall and soil texture', *Bull. Min. Agric. Sudan*, no. 4.
[4] VIDAL-HALL, M. P. (1952) 'The sylviculture and regeneration of the forest types of Equatoria and Bahr el Ghazal Provinces', *Min. Agric. Sudan Forestry Mem.*, no. 4.
[5] DAWKINS, H. C. (1954) 'Notes on a visit to the Imatong Mountains: III. The Imatong Lowland Forest', *Uganda For. Dept. Tech. Note*, no. 6 (Entebbe; roneoed).
[6] ANDREWS, F. W. (1950, 1952) *The Flowering Plants of the Anglo-Egyptian Sudan*, vols I and II (Arbroath).
[7] HEDBERG, O. (1951) 'Vegetation belts of the East African mountains', *Svensk. bot. Tidskr.*, XLV 140.
[8] THOMAS, A. S. (1943) 'The vegetation of Karamoja District, Uganda', *J. Ecol.*, XXXI 149.
[9] EGGELING, W. J. (1947) 'Observations on the ecology of the Budongo rain forest', *J. Ecol.*, XXXIV 20.
[10] JACKSON, J. K. (1950) 'The Dongotona Hills, Sudan', *Emp. For. Rev.*, XXIX 139.
[11] —— (1951) 'Mount Lotuke, Didinga Hills', *Sudan Notes*, XXXII 339.
[12] LUGARD, E. J. H. (1933) 'The flora of Mount Elgon', *Kew Bull.*, I 49.
[13] EDWARDS, D. C., and BOGDAN, A. V. (1951) *Important Grassland Plants of Kenya* (Nairobi).
[14] DALE, I. R. (1936) *Trees and Shrubs of Kenya Colony* (Nairobi).
[15] PITT-SCHENKEL, C. J. W. (1938) 'Some important communities of warm temperate rain forest at Magamba', *J. Ecol.*, XXVI 50.
[16] LEBRUN, J. (1935) 'Les essences forestières des régions montagneuses du Congo Orientale', *Publ. Inst. nat. Agron. Congo Belge, Série scientifique*, no. 1.
[17] SNOWDEN, J. D. (1933) 'A study in altitudinal zonation in South Kigezi and on Mounts Muhavura and Mgahinga, Uganda', *J. Ecol.*, XXI 7.

7 The Subtropical and Subantarctic Rain Forests of New Zealand

L. COCKAYNE

From *The Vegetation of New Zealand*, 2nd ed.; vol. XIV of *Die Vegetation der Erde*, ed. A. Engler and O. Drude (Verlag von Wilhelm Engelmann, Leipzig, 1928) pp. 153–74, 182–6.

PRINCIPLES UPON WHICH THE CLASSIFICATION OF NEW ZEALAND FORESTS IS BASED

THE primary classification of the forest communities into *subtropical* and *subantarctic* rain forest, suggested by me [1], at first thought seems sound enough. But there are wide areas covered by *mixed* forest where podocarps, the usual broad-leaved dicotylous trees and *Nothofagus* are present in abundance, and where the forest interior is almost, if not quite, as hygrophytic as that of subtropical rain forest. It also seems easy to separate the forests on altitudinal lines, but here again a difficulty arises, for certain North Island high-mountain forests, extending from 900 m. to 1200 m. altitude, are but little different from some of those of the lowland Fiord district.

It is the secondary divisions which are the main stumbling-block. Naturally two important points are their floristic composition and ecological features. The former certainly depends largely on latitude and local endemism but, ecologically, there is little difference between a North Island and Stewart Island forest; so, too, with their structure. Two features, however, stand out for subtropical rain forest: that is, it is composed so far as trees go partly of Podocarpaceae, Cupressaceae or Araucariaceae, and partly of broad-leaved dicotylous trees. Unfortunately, for purposes of classification, there is every intermediate stage between a pure podocarp or kauri community and one made up of dicotylous trees alone. Again, a genetic classification suggests itself, but here, too, intermediate stages occur intercalated in one and the same piece of forest.

In what follows, latitudinal and altitudinal distribution of species, structure, special ecology and life-histories have all been taken into consideration. It would have been easy to have made many subdivisions, but I have carefully avoided doing so. The subject is treated rather in a general manner, the main object having been to attempt the presentation of an accurate picture of the remarkable New Zealand rain forest unblurred by superfluous details.

SUBTROPICAL RAIN FOREST OF BROAD-LEAVED
DICOTYLOUS TREES AND CONIFERS

Subtropical rain-forest associations possess so many features in common that, in order to avoid repetition, the following details are submitted.

The trees, shrubs and ferns, with a few trifling exceptions, are evergreen. As viewed from without, the evergreen character of the trees and the general absence of bright greens gives, when seen from a distance, a sombre aspect to the forest, while the density of their growth altogether masks the height of the trees. But a closer view reveals the varied greens and it is not difficult in some instances to recognise certain species from their colour alone, especially in low, even forests of dry ground. An outside view, too, reveals but little of the tropical character of the forest. A few tree-ferns may raise their crowns of spreading, feathery leaves above the greenery, or in North Island and the Sounds–Nelson and North-western districts nikau palms (*Rhopalostylis sapida*) peep forth, but that is all. But push through the belt of shrubs, or low trees, that may fringe its outskirts, and the vision within will be novel enough to one acquainted only with the temperate forests of the northern hemisphere.

Massive trunks, unbranched for many metres, meet the eye, some covered so thickly with lianes and epiphytes, many of which are ferns and bryophytes, that their bark is invisible. Open spaces are few or wanting. Young trees, shrubs of many kinds and tree-ferns 5 to 10 m. tall, growing in clumps or isolated, closely fill the gaps between the tree-trunks. Rope-like stems of lianes depend from the forest-roof swinging in the air, or lie sprawling upon the ground. The bases of the trees are not infrequently swollen and irregular, while their roots spread far and wide over the surface, at times half buried, or, here and there, arching into the air, and covered with seedling trees and shrubs, ferns of goodly size, mantles of mosses and liverworts, lichens and sheets of pellucid Hymenophyllaceae. Fallen trees, in various stages of decay, lie everywhere, and these too, hidden by a garb of water-holding greenery, are the home of seedlings innumerable. The actual forest-floor is most uneven; rotting logs, fallen branches, raised roots, ferns frequently with short trunks and mounds of humus covered with bryophytes and filmy ferns make walking laborious. Progress generally is considerably retarded, too, not merely by the above-mentioned obstacles, or by the close-growing shrubs or spreading branches, but a coarse network of the almost black, stiff stems of the liliaceous liane, *Rhipogonum scandens*, forms entanglements beneath which one is compelled at times to crawl on hands and knees. In other places where there is a good deal of light, the hooked prickles on the midribs of *Rubus australis*, catching a garment, may hold one fast. Furthermore, even on level ground, there are watercourses, here and there, and near these the density of the undergrowth increases, lateral branches from the trees on either side meet and become entangled, the growth of ferns becomes thicker, so that progress is wellnigh impossible. In hilly forest, the density of gullies is still more intensified.

The close growth, which I have attempted to describe, is in harmony with the moist, equable climate, but regulated by the density of the forest-roof. Everywhere is the effect of that complex of factors evoked by the forest itself manifest in the plant-forms. Shrubs, which in the open would be rounded and symmetrical, put forth long, slender stems, that, liane-like, lean against other trees and gain support. Young trees have frequently much-reduced lateral branches and long, straight, slender main-stems. In some this habit is hereditary, and thus the curious juvenile form of *Pseudopanax crassifolium* var. *unifolialatum* may be an 'adaptation' to the forest life.

On the trunks of most of the trees that do not shed their bark in great flakes, and right up on the highest branches, are not only an abundance of true lianes and epiphytes, but seedlings of trees and shrubs, ground ferns of many species and hosts of mosses and liverworts. The trunks of tree-ferns, too, are a favourite station for many plants. Even the slender branches of shrubs may be deeply moss-covered, while leaves themselves may be the home of various small bryophytes.

Certain forests are not nearly so dense or hygrophytic as described above, nor do they exhibit so fully the various peculiarities as cited, but such occupy drier ground than usual, and even then show unmistakably their tropical facies.

In every type of New Zealand rain forest the vegetation is in several distinct layers (stories), each with a definite light-relation. Where the tallest trees are present the uppermost layer (story) consists of their crowns, in some places those of the Podocarpaceae or, in certain associations, *Metrosideros robusta*, one or other of the two species of *Weinmannia*, *Beilschmiedia tawa*, *B. taraire* or species of *Nothofagus*. Trees of a medium size form the next layer, while growing in their crowns, as also in those of the upper story, are the flowering parts of the lianes and in forests of North Island type the more massive epiphytes (e.g. species of *Astelia*, *Griselinia lucida*, *Pittosporum cornifolium*, young *Metrosideros robusta*). The upper layer does not, as a rule, make a continuous roof, so the light-relation, but not the wind-relation, of these two highest stories is not very different. The third layer is formed by the smaller trees, tallest shrubs and tall tree-ferns, while in many North Island forests and the Sounds–Nelson and North-western districts the nikau-palm is conspicuous. Next comes the fourth layer, consisting of the smaller ferns, prostrate or low shrubs, decumbent lianes, tussocks of *Gahnia* or *Astelia* and young plants of various kinds. Finally, there is the layer of actual floor-plants, which is largely made up of small ferns (mostly Hymenophyllaceae) and bryophytes, important genera of which are: (Hepaticae) *Aneura*, *Symphyogyna*, *Monoclea*, *Treubia*, *Chiloscyphus*, *Frullania*, *Mastigobryum*, *Lepidozia*, *Schistochila*, *Trichocolea*, *Plagiochila*, *Tylimanthus*, *Madotheca*; (Musci) *Leucoloma*, *Dicranoloma*, *Leucobryum*, *Leptostomum*, *Hymenodon*, *Bryum*, *Echinodium*, *Ptychomnion*, *Weymouthia*, *Mniodendron*, *Sciadocladus*, *Lembophyllum*, *Distichophyllum*, *Hypopterygium*, *Cyathophorum*, *Rhacopilum*, *Mniadelphus*.

Kauri (Agathis australis) – *broad-leaved dicotylous-tree forest*
Forest of this class is distinguished by the presence of more or less *Agathis australis* together with certain other tall trees, especially *Beilschmiedia tawa*, *B. taraire* (one or both), *Weinmannia sylvicola*, *Metrosideros robusta* and some (or all) of the usual podocarps. Certain species, rare or wanting elsewhere, except in the Auckland districts, are present, of which (excluding those belonging particularly to the kauri subassociation), the following are common or characteristic: *Blechnum fraseri*, *Lygodium articulatum*, *Dacrydium kirkii*, *Mida salicifolia*, *M. myrtifolia*, the hybrids between these two, *Litsaea calicaris*, *Melicytus macrophyllus*, *Dracophyllum latifolium* and the polymorphic *Alseuosmiae*.

The number of species belonging to the community is 231 (pteridophytes 67, spermophytes 164) which belong to 55 families and 117 genera. Lists of the most important species are given when dealing with the associations, etc.

The high value of the timber, together with the inflammability of damaged (*not virgin!*) kauri forest, has led to an enormous reduction in area of the community, so that, except where reserved, a dozen or so years from now will see it virtually gone for ever. But, extensive as were these forests of present-day New Zealand, they were but the remains or successors of more ancient communities, their site plainly marked by abundance of kauri-resin or in places tree-trunks beneath the surface of the ground. Why these ancient forests vanished none can say, but Charles Darwin, who visited a kauri forest in 1835, writes in his *Voyage of the Beagle* in regard to the fernlands near the Bay of Islands, 'Some of the residents think that all this extensive open country originally was covered with forests, and that it has been cleared by fire.' Probably this surmise is true enough for certain localities, but it can hardly be accepted as a general principle.

The forest area, before the interference of man, extended from a line joining Doubtless and Ahipara Bays in the north to the Auckland Isthmus in the south, together with the Barrier Islands and the lower slopes of the Thames Mountains. Kauri forest is essentially an association of the lowlands and lower hills and generally does not ascend to much over 400 m. Nor can it tolerate wet ground, so that its abundant remains in lowland bogs indicates change of level in the land-surface [2].

Partly because of its great monetary value and partly because of the dominating appearance of the lordly kauri, the community, no matter its composition, is generally designated 'kauri forest'. This, however, is a misnomer, for the kauri itself occurs, either as solitary individuals or as large or quite small groups, situated in the general forest-mass, but clearly defined through the presence of the mighty tree unlike any other and the astonishing uniformity in composition of its associated plants.

Although there are changes in regard to latitude, and many differences in the relative abundance of species, the whole kauri-forest community may be considered one association and may be conveniently divided into several subassociations.

As a rule the soil occupied by kauri forest is of poor quality from the agricultural standpoint, a fact emphasised by its occupation after the forest is removed by second-class or usually pasture of a much worse character. But there are various local differences of soil within most of the forest areas and these are reflected by the vegetation, dominance of kauri indicating the least 'fertile' soil.

The kauri (Agathis australis) *subassociation.* This plant-community is distinguished by the dominance of *Agathis australis* and its remarkable assemblage of associated species. The number of kauri trees present range from a close growth of such to isolated trees here and there. The associated species are as follows: *Cyathea dealbata, Dicksonia lanata* (sometimes extremely common), *Blechnum fraseri, Gahnia xanthocarpa, Astelia trinervia, Freycinetia banksii, Mida salicifolia, M. myrtifolia, Weinmannia sylvicola* (juvenile), *Phebalium nudum, Dysoxylum spectabile* (juvenile), *Metrosideros scandens, M. albiflora, Nothopanax arboreum, Dracophyllum latifolium, Leucopogon fasciculatus, Geniostoma ligustrifolium, Coprosma grandifolia, Alseuosmia macrophylla* and *Senecio kirkii.*

The subassociation owes its very characteristic physiognomy partly to the dense tussock-thickets of *Gahnia-Astelia* and partly to the form of the kauri itself. Where it extends over a wide area, the undergrowth is not thick. The kauri-trunks, usually shining-grey, but sometimes reddish, rise up on all sides, as far as the eye can pierce, as massive columns 1 to 3 m. in diameter, unbranched for 20 m. or more (Plate 7). Round the base of each tree is a mound of humus, formed from the shed bark, occupied by small tussocks of *Astelia trinervia,* sprawling *Metrosideros scandens* and *Senecio kirkii.* Rising up between the giant trunks may be multitudes of the straight stems of *Beilschmiedia taraire* thrusting their sparse heads of foliage up to the lower branches of the kauris, or *B. tawa* with its irregular thicker trunk, bryophyte-covered, may be abundant. Between the trees there will be a rather low, open undergrowth consisting of (the first three species being dominant): *Cyathea dealbata* (trunks 1 m. or less), *Dicksonia lanata* (25 cm. high, in colonies, the green fronds arching outwards, their blackish stems shining with a metallic lustre), juvenile *Weinmannia sylvicola* with yellowish-green pinnate leaves, black-stemmed *Wintera axillaris,* slender *Dysoxylum,* graceful young *Beilschmiedia tawa, Melicytus macrophyllus, Suttonia salicina, Coprosma grandifolia,* stemless *Rhopalostylis, Alseuosmia macrophylla, A. quercifolia, A. banksii, A. linariifolia,* if they be species, and the hybrid swarm in which all take a part, and juvenile *Podocarpus ferrugineus.* On the ground will be trailing *Freycinetia,* straggling *Lygodium* (also winding round the young trees), extensive colonies of *Blechnum fraseri*, and in some localities the low straggling shrubs *Pittosporum pimeleoides*, *P. reflexum* and their hybrids, and the fern *Schizaea dichotoma*. High above all rise up the mighty spreading limbs of the kauris, and extending to these lace-like foliage of *Beilschmiedia tawa*, or the darker, denser heads of *B. taraire*. Lianes are not numerous; an occasional *Freycinetia* or *Lygodium* ascend the *Beilschmiedia* trees, but the kauri itself, owing to its bark-shedding habit, remains inviolate.

Much more common than such extensive kauri communities are groves, large or small, or solitary trees, scattered through the forest mass. The kauri trees generally are 20 m. or more distant and the intervening space is occupied by an extremely thick *Gahnia–Astelia* thicket containing also *Freycinetia*, *Metrosideros scandens*, *M. albiflora*, *Phebalium nudum*, *Senecio kirkii* and all the other species already cited as prominent members of the subassociation. Obviously these shrubs, etc., are wanting where the *Gahnia–Astelia* is densest.

The taraire (Beilschmiedia taraire) *subassociation*. In this subassociation the kauri is absent and *B. taraire* dominant. It occurs in its full development to the north of lat. 36° S. The *taraire* trees are about 15 m. high, 3 m. or so apart and their crowns are small but dense. Generally the roof is fairly open. *Metrosideros robusta* is common and as usual conspicuous through its most irregular trunk full of hollows filled with humus supporting so many bryophytes, filmy ferns, herbaceous ferns and shrubs that its brownish, furrowed bark is hidden by the wealth of greenery. Frequently, the trunk leans out of the perpendicular in which case it will support a dense growth of immense asteliads right to the forks of its great twisted branches in the forest-roof. Other common trees are: *Podocarpus totara*, *P. ferrugineus*, *Dacrydium cupressinum*, *Phyllocladus trichomanoides*, *Knightia excelsa*, *Dysoxylum spectabile*, *Beilschmiedia tawa* and *Weinmannia sylvicola*. Much rarer, and not in all localities, are the following: *Libocedrus plumosa*, *Dacrydium kirkii*, *D. colensoi* and *Phyllocladus glaucus*.

The undergrowth, in many places, is not dense, but it varies greatly in this regard according to the intensity of the light. In many parts, it consists of a more or less thick growth of the following young forest-trees, low trees, shrubs and tree-ferns: *Knightia*, *Dysoxylum*, *Beilschmiedia taraire* (these three with a long, straight main-stem and few lateral branches), *Wintera axillaris*, *Pittosporum tenuifolium*,

Melicytus macrophyllus, Myrtus bullata, Nothopanax arboreum, Pseudopanax crassifolium var. *unifoliolatum* or occasionally var. *trifoliolatum, Suttonia australis, S. salicina, Geniostoma ligustrifolium, Coprosma grandifolia, C. arborea* (in dry ground), *Alseuosmia macrophylla, Olearia rani, Cyathea medullaris* and *C. dealbata*. In other parts, there is a thicket of tussocks of *Gahnia xanthocarpa, G. setifolia* and *Astelia trinervia*, mixed with entangled stems of *Rhipogonum scandens* and *Freycinetia banksii*.

The palm, *Rhopalostylis sapida*, is often abundant, its life-form rendering it specially conspicuous. Tree-trunks are draped thickly by climbing species of *Metrosideros* or the *fern Blechnum filiforme*; shrubs and slender trees are festooned and bound together by the twining leaf-spindles of *Lygodium articulatum*; on horizontal branches are masses of *Astelia solandri* and long tassels of *Lycopodium billardieri* swing in the air. *Asplenium flaccidum, A. adiantoides, Dendrobium cunninghami, Pittosporum cornifolium* and the thick-leaved *P. kirkii* are also common epiphytes.

Open spaces of the forest-floor are occupied by mats of vivid-green *Hymenophyllum demissum, Freycinetia*, juvenile *Blechnum filiforme*, rooting *Metrosideros hypericifolia*, seedlings of trees and shrubs and various bryophytes. Fallen trees are thickly covered with translucent *Hymenophyllum dilatatum*, its fronds 30 cm. long, and *Trichomanes reniforme*, dark green when old, but emerald when young.

The tawa (Beilschmiedia tawa) *subassociation*. This subassociation is distinguished by the dominance of *Beilschmiedia tawa*. It occurs to some extent along with the last-described community but, south of lat. 36°, it is the leading broad-leaved tree association. At one time it was dominant on the Waitakerei Hills and Thames Mountains, but in these localities the primitive forest has been in large measure destroyed while the remnant is usually greatly modified.

Thanks to an early paper of Cheeseman's [3], it is possible to give some account of the original vast forest on the Waitakerei Hills since he examined a portion of it before it had been seriously interfered with.

The dominant tree was *Beilschmiedia tawa* which 'probably formed three-fifths of the forest'. The other most abundant trees were *Agathis, Dacrydium cupressinum, Elaeocarpus dentatus, Knightia, Litsaea calicaris, Metrosideros robusta, Pittosporum tenuifolium* and *Suttonia australis*. The undergrowth was dense and consisted of

Alseuosmia macrophylla (abundant), species of *Gahnia, Astelia* and *Coprosma, Rhipogonum, Myrtus bullata* and *Senecio kirkii*; Hymenophyllaceae, ferns in general, and bryophytes were very plentiful.

The originally extensive forests of the Thames subdistrict, except on the Little Barrier Island, Mt Te Aroha and a few other localities, are altogether gone or much modified. Some light is thrown on the original plant-covering by the writings of T. Kirk [4] and J. Adams [5, 6]. Without going into details, the forest was much as already described, but *Beilschmiedia taraire* and *Dicksonia lanata* were scarce.

Pukatea (Laurelia novae-zelandiae)–*nikau* (Rhopalostylis sapida) *subassociation*. What follows refers only to the Waipoua forest (south of Hokianga Harbour, N.A.), but doubtless somewhat similar combinations occurred throughout the plant-association.

The subassociation is confined to moist gulleys. *Beilschmiedia taraire* is rare or absent and *Laurelia novae-zelandiae* the common tree; the palm, generally trunkless, may be so abundant as to dominate. The following are characteristic: *Dicksonia squarrosa* (tree-fern), *Dryopteris pennigera* (here with a trunk), *Asplenium bulbiferum*, mats of *Hymenophyllum demissum*, *Rhipogonum*, *Elatostema rugosum* (the succulent stems with their bronzy leaves raised 1 m. above the ground and occupying many square metres), creeping juvenile *Rubus schmidelioides* and the araliad shrub or tree *Schefflera digitata*. On shaded banks of streams is *Trichomanes rigidum*, its dark fronds covered with small epiphytic mosses.

Life-history of Agathis–*dicotylous forest*. To trace the beginnings of 'kauri forest' the outskirts of the association must be studied where the latter abuts on the contiguous *Leptospermum* shrubland, or the *Pteridium* fernland. Information is also to be procured where regeneration or reinstatement is in progress, the kauri forest having been cut down or burnt. In order to find out the successions which the forest undergoes up to its so-called 'climax', it is necessary to investigate its interior carefully, and to ascertain the light-demanding and shade-enduring capacity of the principal species. As regards the four leading trees, *Agathis* is strongly light-demanding, the two species of *Beilschmiedia* are strongly shade-tolerating and *Weinmannia sylvicola* grows vigorously both in sun and shade.

In many places where the forest and the *Leptospermum* shrubland meet, there is a temporary association (the primary succession), made up of certain shrubland and forest species, including seedling and sapling *Agathis*, the composition of the association being somewhat as follows: *Cyathea dealbata, Blechnum fraseri, Loxsoma cunninghamii* (local), *Lycopodium densum, L. volubile, Gahnia xanthocarpa* (juvenile), *Podocarpus totara, Dacrydium cupressinum, Phyllocladus trichomanoides, Persoonia toru* (specially characteristic), *Knightia excelsa, Weinmannia sylvicola* (juvenile), *Melicytus ramiflorus, Leptospermum scoparium* (dominant), *L. ericoides* (occasionally subdominant), *Nothopanax arboreum, Leucopogon fasciculatus, Geniostoma ligustrifolium, Coprosma robusta, Olearia rani, Brachyglottis repanda* and *Senecio kirkii*. An interesting feature of this transitional forest is the presence of young kauri and young podocarps in much greater abundance than these occur in the forest interior.

Where kauri forest has been destroyed on the Waitakerei Hills it is frequently succeeded by *Leptospermum scoparium* which, after it has attained a considerable size, lets in sufficient light for light-requiring seedlings to gain a footing. The succession developing below the *Leptospermum* is much as already given. *Phyllocladus trichomanoides* is the commonest seedling tree, but there is plenty of *Agathis*. Certain floor-plants are antagonistic to the settlement of trees, e.g. *Schoenus brevifolius, Gleichenia microphylla, Blechnum procerum, B. fraseri* and *Lycopodium densum*.

As the *Leptospermum* grows taller, much more light is let in and eventually the podocarps, kauri and other trees overtop it and it is doomed. Next the sapling forest-trees form a more or less close canopy and shade-tolerating species enter the community, e.g. the species of *Beilschmiedia, Dracophyllum latifolium* and various shrubs and ferns.

The forest having matured, beneath its roof-canopy shade-tolerating trees have slowly developed, ready, at any moment, to replace any mature tree which falls. On the other hand, there are no kauri seedlings in the dense forest. The adult kauris after some hundreds of years reach maturity, and by degrees they die and falling are replaced by the then slender trees of *Beilschmiedia taraire* and *B. tawa*, or it may be *Weinmannia sylvicola*, these forming the dominant members of the final succession (climax succession).

Podocarp–broad leaved dicotylous forest of dry ground
This class of forest is one in which broad-leaved dicotylous trees play a leading part in many places, but in others certain podocarps

occur either more or less evenly dotted here and there or make a close subassociation. In other words, the broad-leaved trees form the groundwork of the community within which the podocarps are inserted. In what follows the name of this class is shortened to 'podocarp–dicotylous forest'.

As for the floristic composition of this group of associations, taking its whole range it embraces almost all the 385 forest species, but in North Island and the north and north-west of South Island it is far richer in species than further to the south.

The forest under consideration extends, but not continuously, from the extreme north of North Island to the south of Stewart Island, but in the conception of such forest the subassociations of kauri forest, excepting the kauri subassociation, should properly be included. Leaving these out of consideration, podocarp–dicotylous forest of the Auckland districts is altogether confined – swamp-forest being excluded – to the montane belt. For the rest of North Island, the community originally occupied most of the soil, its continuity being broken only by areas where the edaphic conditions were unfavourable, particularly swamp and poor soil, but in this regard are many exceptions. In South Island, wide areas in the North-western and Fiord districts are occupied by a mixture of the forest under consideration and that where *Nothofagus* dominates. On the other hand there are pure podocarp–dicotylous forests in the Sounds–Nelson district, while in the Western district, from its northern boundary (R. Taramakau) for about 161 km., there are continuous pure forests of this class. On the east of South Island up to the Dividing Range, the rainfall of the North-eastern, Eastern and North Otago districts is not generally sufficient for any class of forest except semi-swamp forest to establish itself naturally, so in these districts forests (here including all classes except the last-named) occur only near the coast where mountainous (Seaward Kaikoura Mountains, Banks Peninsula), or in gullies, or sheltered places, at the base of the foothills (Mt Oxford, Mt Peel, Orari Gorge, Geraldine, Raincliffe, Waimate). In the South Otago district, however, the forest under consideration followed the coastline (in most places it has been destroyed) and extended inland for a considerable distance in the south. Finally, lowland-montane Stewart Island in large part is occupied by the last-mentioned class of forest.

The podocarp communities

It has already been pointed out that the podocarp content of podocarp–dicotylous forest of dry ground varies from an occasional tree
here and there to subassociations or associations of such magnitude
that the term 'podocarp forest' is no misnomer.

The leading podocarps which form more or less pure stands, and
may even make associations, are (1) the rimu (*Dacrydium cupressinum*), (2) the totara (*Podocarpus totara* at low levels, *P. hallii* at
higher levels usually), (3) the matai (*P. spicatus*). Also the miro
(*P. ferrugineus*) and the kahikatea (*P. dacrydioides*) are present but
usually in much smaller numbers.

All the podocarps require a good deal of light for their early development and without such their growth is either extremely slow or
impossible. *P. totara* is the most light-demanding and probably
P. spicatus the least. Excess of wind, and bright sunshine, is unfavourable, but *P. totara* can tolerate far more of either than the other
species. The astonishing number of nodules on their roots may be a
factor in allowing these podocarps to occupy a class of soil not
favourable for other tall trees (see E. H. Wilson [7] and Yeates [8]),
but judging from the distribution of the forest-trees in general this is
hardly likely, and various species (e.g. *Podocarpus nivalis, Dacrydium
bidwillii*) may occupy xerophytic stations.

Rimu (Dacrydium cupressinum) *communities.* These are distinguished
by the dominance of *D. cupressinum*, but other podocarps, and some
broad-leaved trees, are generally present.

The floristic character of the communities changes greatly in
proceeding from north to south, but this is dealt with when treating
of the broad-leaved dicotylous-tree communities.

Rimu forest, if it may be so called, extends from the high land
south of Hokianga Harbour to Stewart Island, but there *Weinmannia racemosa* is present in equal or greater quantity. In the
Auckland districts it does not occur at much below 600 m. altitude
and even in the East Cape and Volcanic Plateau districts it is an
upland community. In the Western district, at the present time, there
is a great deal of rimu forest and the rimu is particularly tall, but
not of excessive girth.

The physiognomy of the forest needs but little description, since
that already given for New Zealand rain forest in general applies
quite well. The one striking and peculiar feature is dependent on

the rimu itself with its long, straight trunk crowned by a rather small yellowish-green head of drooping-shoots.

Totara (Podocarpus totara, P. hallii – *at times*) *communities.* 'Totara forest' is podocarp forest in which either *P. totara* or *P. hallii*, or both, are dominant. The association is more xerophytic than that of *Dacrydium cupressinum*, the species being able to occupy dry ground or exposed positions where the latter would perish. At the same time, it must be pointed out that totara to a varying extent is nearly always present in podocarp–dictoylous forest and may equal the rimu in importance, in which case there would be a rimu–totara association.

At the present time it is not possible to state accurately the distribution of those forests where totara dominated or was present in abundance. Campbell-Walker [9] in his account of forest distribution in New Zealand defined a 'central or totara district', which included all the forest-lands of the East Cape, Volcanic Plateau and Ruahine–Cook districts. Now although his area as a whole certainly contained abundant totara (*Podocarpus totara*), yet there were also rimu, kahikatea and other forest communities and it is probable that the actual totara associations or subassociations were limited, as now, to the Volcanic Plateau from the north and west of Lake Taupo to the main-trunk line and to portions of the East Cape district. Elsewhere there were most likely rimu–totara and rimu–totara–matai associations. In South Island totara forest appears to have been almost restricted to the Eastern district, although both species are of wide distribution and certainly occurred in considerable quantities in all the districts, except North Otago, while in the Western *P. hallii* is the dominant species of the lower subalpine forest and in Stewart Island the only one.

At one time totara forest was much more widespread in the Eastern and North Otago districts than is now the case, but how long ago this was who can say? All we know is that totara logs lay on the ground in abundance in Central Otago and parts of Canterbury now treeless. There is no clue as to what caused the destruction of these ancient forests. The story goes that there was a vast forest-fire in pre-European days, but it is almost impossible to see how this could cause such wholesale and absolute destruction or why the fallen logs remained. Still more remarkable is the fact of a forest, still undestroyed, marking the limit of the western rainfall. The climate, too, where those tree-remains lie, is distinctly too dry for the *natural* occupation by totara forest and I can only conclude with Speight [10] that the forest came into existence during a much wetter period than the present. This also would account for the rarity of *Dacrydium cupressinum* on Banks Peninsula, the wet post-glacial period leading to the replacement of a primitive *Nothofagus* forest by one with *D. cupressinum* dominant, but the latter during a subsequent drier period, in its turn, being replaced by *Podocarpus totara*.

Wherever totara forest is situated, its composition is similar to that of other adjacent rain-forest associations. A brief description of the association (now destroyed) at an altitude of some 300 m. near Taumarunui will give some idea of the forest as it occurred in the most extensive area of its distribution.

The soil was pumice mixed near the surface with a good deal of humus. Besides the dominant *P. totara* there was much *P. spicatus*. The podocarps formed the upper tier of foliage; their straight, columnar trunks were a striking feature, those of *P. totara* bearing but a scanty covering of bryophytes, its outer bark hanging in long strips. Some of the other forest trees were *Knightia excelsa*, *Beilschmiedia tawa* (abundant), *Carpodetus serratus*, *Weinmannia racemosa*, *Pennantia corymbosa*, *Alectryon excelsum*, *Hoheria sexstylosa*, *Melicytus ramiflorus*, *Fuchsia excorticata*, *Nothopanax arboreum*, *Suttonia australis*, *Olea montana* and *Brachyglottis repanda*. Most of the above also occurred as shrubs or young trees of undergrowth in which likewise, amongst others, were the following: *Paratrophis microphylla* (very common), *Aristotelia serrata*, *Myrtus pedunculata*, *Schefflera digitata* (dominant in some places), *Coprosma rotundifolia* and *Rhabdothamnus solandri*. In some gullies of this particular community were hundreds of young plants of *Schefflera* 30 cm. or so high, associated with various ground-ferns, especially *Dryopteris pennigera*. The chief tree-fern, liane and epiphyte respectively were *Cyathea dealbata*, *Metrosideros hypericifolia* and *Astelia solandri*. Bryophytes and Hymenophyllaceae were plentiful.

The forest of Banks Peninsula, now almost gone, originally consisted largely of a *Podocarpus totara* association, but in places *P. spicatus* was dominant, except on certain slopes facing north and in the subalpine belt. In the deeper gullies, where there were permanent streams, was a wealth of ferns including Hymenophyllaceae, but according to Armstrong's list [11], and to recent observations, only two or three species of the latter were at all plentiful. On many slopes and in the waterless stony gullies, the forest was of a dry character, as evidenced by the abundance of *Pellaea rotundifolia* and *Polystichum richardi.* Certain negative features help to define the forest as a whole. Thus the following, some of which extend much further to the south in the west of the island, were absent: *Trichomanes reniforme, Lindsaya cuneata, Freycinetia banksii, Astelia cunninghamii, Ascarina lucida, Weinmannia racemosa, Metrosideros lucida, Metrosideros scandens, Nothopanax edgerleyi* and *Suttonia salicina,* while *Rhopalostylis,* though present in a few places, was generally wanting. On the other hand, the elsewhere rare liane *Senecio sciadophilus* is still abundant, *Tetrapathaea tetrandra* is not uncommon, the rare *Teucridium parvifolium* is plentiful, the semi-liane *Microlaena polynoda* is fairly frequent and *Cyathea medullaris, Australina pusilla, Corynocarpus laevigata, Pseudopanax ferox* and *Olearia fragrantissima* were occasionally present.

Matai (Podocarpus spicatus) *communities.* To what extent *Podocarpus spicatus* formed communities, more or less pure throughout its range, it is now impossible to state. It is rare in many forest areas and, even where common, hardly forms even a colony.

A portion of the Mt Peel podocarp forest seems, according to H. H. Allan [12], to be an exception to the above statement. Thus for the forest of the upper terrace flats he writes, 'In general *P. spicatus* is dominant with *P. totara* and *P. dacrydioides* in lesser amounts'. On the driest ground *P. totara* dominates. Also, even yet, a *P. spicatus* association exists on certain parts of the Southland Plain, the trees close together, rather stunted and their crowns more spreading than usual. According to Roberts, *P. spicatus* forms small societies on the flats of nearly all the Westland rivers as far south as the Cascade river. He mentions also stunted trees, 2·4 m. in diameter with 'short bunched trunks dividing into several long, heavy branches'.

P. spicatus, now extremely rare in Stewart Island, must have been dominant or subdominant, for the remains of trees, probably dead for 300 to 400 years, lie abundantly upon the forest-floor, but frequently embraced by the trunk, originally the root, of a mature *Weinmannia racemosa.*

The broad-leaved dicotylous tree communities
The forests here dealt with are those occasionally composed almost entirely of broad-leaved dicotylous trees, but generally with more

or less podocarps present, which usually play a minor part, so far as tall trees are concerned, but, as already shown, they may make communities of different grades within the general forest mass or even compose the greater part of the latter; or again, podocarps and dicotylous trees may be about equal in number. There are, indeed, no hard and fast lines, so that much of the classification proposed by anyone can hardly fail to be of an artificial character, or it may be detailed to excess and useless for practical purposes.

The forest communities under consideration, unlike those of kauri or podocarp, are composed of dominant trees which when young are either shade-tolerating or epiphytic, so that, under certain circumstances, they are able to replace the conifers. Thus, the forests they eventually govern are climax successions.

Obviously the distribution of these broad-leaved dicotylous forests is that of lowland-montane forest in general as already described. At one time they occupied an area far greater than now, owing to their having been destroyed, year by year, over wide areas to give place to artificial grassland, whereas forests containing an abundance of podocarps – swamp-forest excepted – are removed much more slowly through saw-milling and their damaged remnants may persist for many years.

Unlike the podocarps, which thrive equally well from the north of North Island to Stewart Island, the dominant broad-leaved trees differ in their frost-tolerating capacity, so that the communities are climatic.

Tawa (Beilschmiedia tawa) *communities.* The dominant tree is *B. tawa*. The undergrowth and species are those of subtropical rain forest of North Island and the Sounds–Nelson district.

It extends from the north of North Island to the Sounds–Nelson district. In North Island it ascends to 600 m. or more. Cook Strait notwithstanding, the forests on both sides of this apparently natural obstacle to distribution are of similar composition – a minor distinction being the occasional presence in Sounds–Nelson of the herbs *Poranthera microphylla* and *Scutellaria novae-zelandiae*.

Frequently, but in the southern part of its range more especially, tawa forest comes into contact with *Nothofagus* forest, but the latter in general is confined to the more barren, dry slopes or ridges, while the tawa is in the gullies and the rich alluvial soil of the flat ground.

Seen from a distance, the tawa trees are of a rather unpleasing grey colour; nevertheless, at a close view with the crowns of tree-ferns peeping out of the dense undergrowth with its vivid greens, the scene is pleasing enough.

Unlike the outskirts of kauri or podocarp–dicotylous forest and *Nothofagus* forest, few seedlings of the dominant tree are to be seen there, but under the forest-roof or the canopy of small trees of the next story, in open places there will be seedlings in their hundreds; tawa saplings, too, are a frequent feature of the undergrowth.

Montane tawa forest in the Auckland and East Cape districts contains a good deal of the monotypic *Ixerba brexioides*[1] and a fair amount of *Quintinia serrata*[2] – both trees of physiognomic import-ance. On the Mamaku Plateau *Weinmannia racemosa* (not *W. sylvicola* as further north) is a common tree, *Dicksonia fibrosa* is an abundant tree-fern and *Alseuosmia macrophylla* is frequent in the undergrowth.

Kamahi (Weinmannia racemosa) *communities*. These are distinguished by the dominance of *W. racemosa*, or there may be a good deal of *Dacrydium cupressinum* dotted about or in colonies, while there are almost always some of the other podocarps.

Weinmannia forest is of wide distribution and extends from the Mamaku Plateau in the north to Stewart Island, thanks to the frost-tolerating capacity of the tree and its ready establishment. Thus it rarely commences life as a terrestial plant, but begins as an epiphyte upon a tree-fern stem, or as a seedling upon some fallen moss-covered trunk. The epiphytic habit is most advantageous in dense forest, for not only does the seedling escape competition with the floor plants, but is in a better position than most with regard to light. As the seedling grows into a tree its roots embrace its host or the fallen tree and eventually, growing together, function as the base of the trunk. Probably *Weinmannia racemosa* is the commonest massive forest-tree in New Zealand, though most who work in the forest would select *Dacrydium cupressinum* for that honour, but rather by reason of its being a timber-tree with which they are specially concerned than because of its relative abundance in forests

---

[1] A beautiful shade-demanding tree of slow growth, 6 to 12 m. high with dark-green, glossy, serrate, narrow-lanceolate leaves ± 12 cm. long, arranged in whorls near the apices of the branches and bearing in early summer abundance of white flowers about 2·5 cm. in diameter arranged in terminal panicles.

[2] A rather smaller tree than *Ixerba* with pale, rather thin, greenish-yellow, narrow-oblong leaves ± 12 cm. long and their margins crinkled.

generally. Frequently, valuable so-called 'rimu' milling-forest contains twice as much or considerably more kamahi than rimu.

Weinmannia racemosa forest is both lowland and montane, the species itself ascending into the subalpine belt. As for its composition, that of North Island, and the north-east of South Island (S.N.), is similar to that of the podocarp–dicotylous forests already dealt with. But in the North-western and Western districts, certain species rare or wanting elsewhere enter in, particularly *Ascarina lucida*[1] and *Quintinia acutifolia*,[2] as also much juvenile *Elaeocarpus hookerianus* of divaricating form.[3] Generally there is more or less *Dacrydium cupressinum* or indeed it may dominate. From any lowland-forest association of the eastern and southern parts of South Island and from that of Stewart Island the association is distinguished by the presence in abundance of the lianes *Freycinetia* and *Metrosideros scandens* and the epiphyte *Astelia cunninghamii* and by the absence of *Pittosporum eugenioides* and *P. tenuifolium*. Byophytes, though not building cushions, are abundant enough to be of prime physiognomic importance, especially *Weymouthia mollis* and the larger *W. billardieri* hanging from slender branches or twigs and, on the forest-floor, *Plagiochila gigantea* and other species of the genus, extensive mats of the pale-green *Trichocolea tomentella* frequently glistening with drops of water and species of *Schistochila*. Ferns are extremely abundant on the forest-floor, including colonies of *Gleichenia cunninghamii*, abundant *Leptopteris superba* and *Blechnum nigrum* (in the darkest places).

In the South Otago and Stewart districts, the *Dacrydium–Weinmannia* forest is much the same for both districts, so that one description will suffice for the two. Generally some *Metrosideros lucida* is present and at times in such abundance, especially near the sea or on hillsides, that it equals the other two trees.

The dominant shrub of the undergrowth is *Coprosma foetidissima*; *C. colensoi* and *C. astoni* are common as also the hybrids between the three. *Rhipogonum* is the only important liane. Asteliads, as epiphytes,

---

[1] *Ascarina lucida* is a low, bushy tree with almost black bark and green, extremely glossy, oblong serrate leaves 2·5 to 5 cm. long.

[2] *Quintinia acutifolia* is a small, rather fastigiate tree of slender habit and abundant leaves of oblong type 7·5 to 12 cm. long, yellowish with green veins and midrib. The flowers are pale lilac and arranged in many-flowered racemes about 10 cm. long.

[3] Other important shrubs, etc., of the undergrowth are small *Podocarpus ferrugineus*, *Carpodetus*, juvenile *Pseudopanax crassifolium* var. *unifoliolatum*, *Schefflera digitata* and *Coprosma foetidissima*.

are absent but *Dendrobium cunninghamii* is still plentiful and *Griselinia littoralis* replaces its epiphytic relative of the north. *Hemitelia* is the prevalent tree-fern, and as ground ferns *Blechnum discolor, B. procerum* and at times *Leptopteris superba* make extensive colonies. In the north of the South Otago district *Weinmannia* was virtually absent in the originally extensive forest area. *Pittosporum tenuifolium, P. eugenoides* and *Nothopanax arboreum* are absent in Stewart Island, the place of the first-named being filled by *P. fasciculatum* and of the last by *N. colensoi.*

Northern-rata (Metrosideros robusta) *forest.* A forest or a minor community of this class is distinguished by *Metrosideros robusta* – that huge tree of most irregular form, owing to its epiphytic origin – its spreading limbs bearing veritable gardens of shrubs, ferns, a pendent lycopod, and one or two great asteliads. Its composition is similar to that of the rimu forest which it has replaced or is replacing, as will be seen below. *Weinmannia racemosa* and *Beilschmiedia tawa* are usually important trees. The base of the northern-rata will be covered with various Hymenophyllaceae (e.g. *Trichomanes reniforme* and *Hymenophyllum flabellatum* in dense mats), bryophytes (including cushions of *Leucobryum candidum*) and many seedling trees and shrubs. Where the forest is montane, or upper lowland, there probably will be a good deal of the beautiful *Senecio kirkii* on the base of the rata, its leaves soft, dark green and fleshy and the snow-white flower-heads, each 3 cm in diameter, in great abundance.

Southern-rata (Metrosideros lucida) *forest.* This is distinguished by the dominance of *M. lucida*, but frequently there is an equal amount of *Weinmannia racemosa*, and such would be 'southern rata–kamahi forest'. The community belongs essentially to South Island, with its chief development in the North-western, Western, South Otago, Stewart and perhaps Fiord districts, being in the first two districts an upper lowland and montane forest and in the others frequently coastal and semi-coastal as well as montane.

In the Western district at above 450 m. altitude *M. lucida* becomes dominant and the lowland podocarps gradually decrease in numbers. *Weinmannia racemosa* is so plentiful in places as to dominate. *Quintinia acutifolia* is conspicuous through its somewhat fastigiate habit as a sapling and the yellowish leaves blotched with purple but pale beneath. Many of the lowland shrubs and ferns are present.

The undergrowth is dense, especially in gullies. Bryophytes (species of *Gottschea*, *Schistochila*, *Aneura*, *Mniodendron*, *Plagiochila*, *Lembophyllum*, etc.) and Hymenophyllaceae abound. *Leptopteris superba* forms extensive colonies.

At the Franz Josef Glacier, the terminal face of which descends to 213 m., the southern-rata association comes on to the ice-worn rocks at a few metres from the ice on either side of the glacier. The forest here, the roof of which has the characteristic billowy appearance, consists principally of the following: *Metrosideros lucida* and *Weinmannia racemosa* (the dominant canopy trees), *Carpodetus serratus*, *Coriaria arborea*, *Aristotelia serrata*, *Hoheria glabrata*, *Melicytus ramiflorus*, *Pseudopanax crassifolium* var. *unifoliolatum*, *Schefflera digitata*, *Griselinia littoralis*, *Hebe salicifolia*, *Coprosma lucida*, *Olearia arborescens* and *O. avicenniaefolia*. The pteridophytes include *Hemitelia smithii* (tree-fern, but here of low stature), several Hymenophyllaceae, *Hypolepis tenuifolia*, *Histiopteris incisa*, *Blechnum procerum*, *B. lanceolatum*, *Asplenium bulbiferum*, *A. flaccidum*, *Polystichum vestitum*, *Polypodium diversifolium*, *P. billardieri* and *Lycopodium volubile*.

In Stewart Island, southern-rata forest, except close to the shores of the inlets, or on small islands therein, is a montane community. In exposed places the trunk may be prostrate or semi-prostrate as in Lord Auckland Island. The undergrowth is that of ordinary Stewart Island forest.

On Mt Peel (E.), according to H. H. Allan [12], *M. lucida* forms a subassociation on 'rocky knolls and slopes with a western aspect', the tree attaining a height of about 7 m. The undergrowth is sparse and is partly made up of *Cyathodes acerosa*, *Coprosma parviflora*, *C. rhamnoides* and *Suttonia australis*.

The life-history of podocarp–dicotylous broad-leaved forest of dry ground

The early beginnings of the forest under consideration have already been explained in the general introduction to forest, so far as my limited knowledge of the subject goes. Also it has been pointed out that it is not possible to give an account of the various successions, for with the exceptions dealt with below, it seems to be rather a gradual process, except in the earlier stages, leading up to a climax than a series of temporary associations terminating in such a climax – but this statement, opposed to accepted ecological teaching, may be

1 *Undergrowth of Mora forest, Moraballi Creek, British Guiana. The buttressed tree is* Mora excelsa, *belonging to the A story. The small tree on the right is* Duguetia sp. *and shows the habit characteristic of trees in the C story. (From P. W. Richards,* The Tropical Rain Forest, *plate II*B.)

2 *Canopy of rain forest, Shasha Forest Reserve, Nigeria, seen from 24 m. above ground. In foreground C story trees bound together into a dense mass by lianes; in background 'emergent' trees of the A and B stories. (From P. W. Richards,* The Tropical Rain Forest, *plate III*A.)

3 *Semi-evergreen seasonal fores* *Mayaro, east of the Nariṭ* *Swamp, Trinidad.* (*From J. S* *Beard*, The Natural Vegetation ɑ Trinidad, *fig. 14.*)

5 Nothofagus *beech forest cover-* *ing rugged topography in th* *Kubor Ranges at the head of th* *Minj River valley, New Guinea* *Tufted* Cordyline *and plume* *heads of swordgrass marked th* *edge of a garden at 8,000 ft* Casuarina *trees may be seen at th* *left in the inhabited valley below* (*From R. G. Robbins, 'Th* *montane vegetation of New* *Guinea'*, Tuatara, VIII 3, p. 127.

4 *Secondary deciduous seasonal forest, Chacachacare, Trinidad. (From J. S. Beard*, The Natural Vegetation of Trinidad, *fig. 17*.)

6 *Tussocks of alpine grassland in the summit area of Mt Hagen, New Guinea, with outliers of montane forest and subalpine shrub at 11,000 ft. Lava outcrops on the outer slopes of the old crater rim. (From R. G. Robbins*, 'The montane vegetation of New Guinea', Tuatara, VIII *3, p. 127*.)

8 *Pine–hemlock forest* (Pinus strobus canadensis) *in Pennsylvania.* (*Photograph, cou̶Forest Service; from J. E. Weaver and F.* Plant Ecology, *fig. 252.*)

7   Interior of an extensive area of the kauri subassociation. The large trunks are those of the kauri (Agathis australis) *and the slender ones mostly those of* Beilschmiedia taraire. *Waipoua Kauri Forest, North Auckland district, in 1907. (From L. Cockayne,* The Vegetation of New Zealand, *fig. 35.)*

10(a)   Chaparral of Adenostoma fasciculatum *and* Ceanothus crassifolius (light), *a common combination in southern California. Near Wheeler Hot Springs, Ventura County. (From W. S. Cooper,* The Broad-Sclerophyll Vegetation of California, *plate 12A.)*

(b)   Quercus agrifolia–Arbutus association on north slope of Jasper Ridge; chaparral on ridge top. (From W. S. Cooper, The Broad-Sclerophyll Vegetation of California, *plate 14A.)*

9   Virgin forest of white pine (Pinus strobus) *and hemlock* (Tsuga canadensis) *in the Hearts Content area of the Allegheny National Forest, Pennsylvania.*

11(a)   *Mixed community of mallee* (Eucalyptus oleosa), *saltbush and bluebush at Dilkera, showing typical habit of the mallee.* (*From J. G. Wood, 'Floristics and ecology of the mallee',* Trans. Roy. Soc. S. Aust., LIII, *fig. 1.*)

(b)   *Meadow steppe on the slope of a steppe valley in the neighbourhood of oak forest.* Stipa joannis, Dactylis glomerata, Filipendula hexapetala, Chrysanthemum leucanthemum. Gov. Vorónezh, District of Usman. (*From Boris Keller, 'Distribution of vegetation on the plains of European Russia',* J. Ecol., XV, *photo 5.*)

12(a) *Young aspen grove, fifteen years old, with border dominated by* Symphoricarpos occidentalis; *prairie vegetation in foreground. (From E. H. Moss, 'The vegetation of Alberta . . .', J. Ecol., xx, photo 6.)*

(b) *Parkland; aspen vegetation on north slopes, prairie on south slopes. (From E. H. Moss, 'The vegetation of Alberta . . .', J. Ecol., xx, photo 7.)*

13 *Saguaro National Monument near Tucson, facing north-east towards Agua Caliente Hill. The photograph, taken in 1960, shows that the saguaro population has undergone a reduction of about one-third since about 1935. A recent study of the stand indicates that, if the present trend continues, the cactus will disappear by 1998 (Alcorn and May, 'Attrition of a saguaro forest',* Plant Disease Reporter, xlvi (1962) 157). *(From J. R. Hastings and R. M. Turner,* The Changing Mile, *plate 61b.)*

14 Dry tundra on raised area overlooking Hudson Bay, composed principally of an intricate mixture of xerophilous lichens, grasses, sedges and other herbs. Dwarf woody plants also occur, and the surface is interrupted by projecting lichen-covered boulders. (*From N. Polunin*, Introduction to Plant Geography, *fig. 111*.)

15 Tangled willow scrub in the low-arctic belt of the Northwest Territories, Canada. The scrub occupies a slight depression whose depth is indicated by a spade on the ground (*centre*). (*From N. Polunin*, Introduction to Plant Geography, *fig. 115*.)

16 'Snow patch' darkened Arctic Bell-heather (Cassic tetragona), *southern Baffin Isla From N. Polunin*, Introduction Plant Geography, *fig. 118*.)

due perhaps to my ignorance rather than to the real facts of the case.

As already seen, the two distinct groups of associations or sub-associations which stand out are those composed respectively of podocarps and broad-leaved dicotylous trees. Owing to the podocarps being light-demanding, they become members of the forest earlier than the bulk of the other trees. Also, theoretically, podocarp forests should have been in existence long before dicotylous trees were evolved. Be this as it may, podocarps, along with certain dicotylous trees and shrubs, are important members of the youngest forest associations (successions). Nothwithstanding their slow growth, and because it is little if any slower than that of the tall or medium-sized dicotylous trees, the podocarps reach a height which cannot be overtaken by the later-arriving dicotylous competitors, e.g. *Beilschmiedia tawa*. Thus, in young forest, there will always have been a strong podocarp element and, in the early history of those forests of which the present are the direct descendants, the podocarps would be supreme. Nor, even yet, are they readily supplanted by their rivals, for most of the species can exist as 'lingerers' for many years, ready to grow with considerable vigour as soon as light is let in to the interior of the forest through the falling of some over-mature tree.

When, in the course of its development, the light within the young forest becomes more subdued, the various shade-tolerating trees, shrubs and ferns put in an appearance. These, so long as the under-growth does not become too dense, even for them, will grow at a fair pace and, as low trees, etc, thrive beneath the forest roof. By degrees, too, with decrease of light, the shade-demanding species will enter the community.

Within the forest there is great competition between the species. Where there is sufficient light, colonies of tree-ferns (these are by no means purely shade-plants) are readily established, and these forbid the presence of seedlings through the dense shade they cast. Favoured by rather dry ground, wide breadths of open forest-floor are rapidly occupied by colonies of *Blechnum discolor* which are hostile to the incoming of *all* seedlings. Certain lianes, particularly *Rhipogonum scandens*, destroy the shrubs and young trees which they embrace and entanglements of naked liane-stems result.

In many North Island forests, hundreds of seedlings of *Beilschmiedia tawa* are to be seen in the more open places, thanks to the high germinating-power of the seeds, the large fruits of which lie where they fall, and the shade-tolerating capacity of the young plants. With but few competitors, the advantage is greatly on the side of the young tawas, some of which grow into saplings; indeed, sapling tawas often form a considerable percentage of the under-growth, while, beneath them, there may be seedling tawas in profusion. *Forest of this class is potential tawa forest.* In fact, every transition can be observed from pure podocarp forest to that where the podocarps are altogether wanting over a wide area, and where nearly all the tall trees are *Beilschmiedia tawa.*

In certain North Island tree communities the tawa is far less in evidence, but the podocarps – particularly *Dacrydium cupressinum* (rimu) – have an openly-declared enemy in *Metrosideros robusta* (northern-rata), which as an epiphyte – thanks to its minute seeds – so readily gains a footing on the boughs of the rimu. Very soon the humus made by the epiphytic asteliads, etc., that is the soil on which the northern-rata's seeds have germinated, becomes insufficient for the rapidly-growing shrub and this puts down roots which, in course of time, reach the ground and eventually crush and kill their host, and growing together form an enormous trunk irregular in shape. This replacement of rimu forest can be seen at every stage of progress in many forest areas up to the southern limit reached by the northern-rata.

Within the forest a similar phenomenon takes place when tree-ferns are attacked by *Nothopanax arboreum* (also the other small trees of this genus) as an epiphyte, and colonies of this small tree are frequent which have originated in this manner.

The establishment of a *Weinmannia racemosa* (kamahi) climax takes place in somewhat the same manner as that of the semi-epiphytes cited above. The tree itself begins life (1) as a seedling upon the ground, in which case it rarely reaches beyond the shrub stage, (2) as an epiphyte on the trunk of a tree-fern, and (3) the seedling develops upon a fallen tree-trunk. It is the last two cases which concern the incoming of the *Weinmannia* climax-association, for in both the light is sufficient for the fairly rapid development of the young tree, both the fallen trunk and the tree-fern indicating an open roof-canopy.

On Banks Peninsula and certain other places where sapling tall trees ready to replace the podocarps are few in numbers, it seems probable that the climax-forest is made up of various shade-demanding or shade-tolerating small trees.

On the slopes of gullies tall trees are absent. There is sufficient illumination for light-demanding species, and such portions of a forest may be considered migratory climaxes. In such, *Fuchsia exorticata* plays an important part; other common members of the community are *Rhipogonum scandens*, *Muehlenbeckia australis*, *Weinmannia racemosa* (as a small bushy tree), *Aristotelia serrata*, *Melicytus ramiflorus* and *Shefflera digitata*.

SUBANTARCTIC RAIN FOREST

The community under consideration is distinguished by the dominance of one or more species of *Nothofagus*, and the rarity of other tall trees. Nevertheless, especially in the west of South Island, there are forests where the subtropical and subantarctic forest trees are equal in number, such communities grading gradually into pure forest of either class, as the case may be.

In North Island and the Sounds–Nelson district (South Island), *Nothofagus truncata* and *N. solandri* dominate, but in the west and south of South Island the dominant trees are *N. menziesii*, *N. fusca* and *N. cliffortioides*, all of which or one only being present. The hybrid swarms × *N. cliffusca* and × *soltruncata* are widely distributed and sometimes dominate small areas.

Subantarctic forest is first met with in the Thames subdistrict, whence it follows the dividing-range of North Island and near Cook Strait occurs almost at sea-level. It is found also to some extent in the broken country to the east of the Egmont–Wanganui district. In South Island an association, almost identical with that of the southern Ruahine–Cook district, occurs in the Sounds–Nelson district and extends to the eastern part of the North-western district. To the west of the Tasman Mountains and the Southern Alps, excepting from the R. Taramakau to the R. Paringa, pure or mixed *Nothofagus* forest extends to the south coast, and eastwards passes for a considerable distance into the South Otago district.[1]

The *Nothofagus* communities may be naturally classified in terms of the dominance of any species of that genus. But a better conception of this class of forest, as a whole, is to be gained as follows from

[1] There are some patches of lowland *Nothofagus* forest in the east of the Northeastern district, also at one time there were such on Banks Peninsula, and even yet such forest exists in a number of places in the neighbourhood of Dunedin (recently made known by J. S. Thomson and Simpson), and the vicinity of Catlins river.

an account of typical portions of the community in proceeding from
north to south.

Montane Nothofagus *forest of the Mamaku Plateau.* This is distinguished by the dominance of *Nothofagus menziesii* and *N. fusca*,
and the presence of the small tree, *Phyllocladus glaucus*; there is also
a little *N. truncata*. *Weinmannia racemosa* dominates in some places
and, in others, *Beilschmiedia tawa* is abundant. *Ixerba brexioides*,
both juvenile and adult, occur in the undergrowth and *Quintinia
serrata* and *Alseuosmia macrophylla* are plentiful. Where the roof-
canopy is open there is abundance of young *Nothofagi*.

Besides the species already cited, the following are common:
*Hymenophyllum scabrum, H. flabellatum, H. multifidum, Tricho-
manes reniforme, Dicksonia squarrosa, Blechnum discolor, Leptopteris superba, Gahnia pauciflora, Astelia nervosa* var. *sylvestris,
Wintera colorata, Wintera axillaris, Carpodetus serratus, Elaeocarpus dentatus, Melicytus lanceolatus, Myrtus pedunculata, Nothopanax arboreum, N. edgerleyi, N. anomalum, Suttonia salicina,
Coprosma grandifolia* and *Senecio kirkii*.

The presence of *Nothofagus* forest on the Mamaku Plateau, together with that of *Phyllocladus glaucus*, according to B. C. Aston,
depends upon the nature of the soil which is a 'sandy loam' distinct
from the 'air-borne sandy silts bearing the typical tawa–rimu forest
of the kind most resistent to climatic severity'.

Nothofagus solandri–truncata *association.* Forest of this class is
distinguished by the dominance of *Nothofagus solandri* and *N.
truncata*. Frequently it grows side by side with dicotylous–podocarp
forest, the latter occupying gullies and flat ground and the former the
ridges or wherever the soil is 'poorest'.

There are fewer species than in the adjacent dicotylous–podocarp
forest. The trees of *N. truncata* generally are of large size – say 24 m.
high, or more, and up to 85 cm. or more in diameter – but if specially
massive they usually are more or less decayed; *N. solandri* is smaller.
The undergrowth is often scanty; it consists of species more tolerant
of a dry habitat than those in general of the adjacent subtropical
rain forest, especially: *Nothopanax arboreum* (of non-epiphytic
origin), *Cyathodes acerosa*, juvenile *Weinmannia racemosa* (of non-
epiphytic origin), *Leucopogon fasciculatus, Geniostoma ligustrifolium, Coprosma rhamnoides*, and, where there is abundant light,

young *Nothofagi*. Where particularly dry, the usually epiphytic *Astelia solandri* and *Earina autumnale* are common on the forest-floor and there are frequently carpets of *Trichomanes reniforme*. *Cyathea dealbata* is the common tree-fern. Lianes and large epiphytes are of little moment. Foliaceous lichens are common on tree-trunks.

In some parts of the class of forest under consideration *Dacrydium cupressinum*, or other podocarps, occur in limited quantity, and there are transitions leading to dicotylous–podocarp forest.

The association just described is that of the southern Ruahine–Cook district and the Sounds–Nelson district. An association similar in character was at one time common at the base of the Ruahine Mountains, and that on sandstone ridges in the Egmont–Wanganui district is probably similar.

Lowland Nothofagus *forest in the west of the North-western district.* The *Nothofagus* forest in this area is by no means uniform in its composition. *N. fusca, N. menziesii, N. cliffortioides* and some *N. truncata* may all be present, or either of the first two be the sole tree, or *N. cliffortioides* be present in about equal quantity.

The associations occur both in river valleys and on hillsides, but on the most 'fertile' soil there is dicotylous–podocarp forest containing more or less *Nothofagus*.

Owing to the wet climate, the undergrowth is similar to that of the neighbouring dicotylous–podocarp forest, the following being characteristic species: *Alsophila colensoi, Polystichum vestitum, Leptopteris superba, Quintinia acutifolia* (but not everywhere), *Wintera colorata, Pittosporum divaricatum* (but not everywhere), *Viola filicaulis, Myrtus pedunculata, Nothopanax simplex, N. anomalum, Suttonia divaricata, Coprosma foetidissima* and *Nertera dichondraefolia* – all, except perhaps the last, being common lower-subalpine species.

On the floor, mats of *N. dichondraefolia* and bryophytes are common. Large foliaceous lichens (species of *Sticta*, etc.) abound on tree-trunks and twigs (e.g. those of *Myrtus pedunculata*). The undergrowth consists of low trees with slender trunks, twiggy shrubs covered with epiphytic bryophytes, occasional tree-ferns (*Hemitelia smithii, Dicksonia squarrosa*), and, according to the light-intensity, more or less sapling *Nothofagi*.

Lowland Nothofagus *forest of the Fiord district.* Generally *Nothofagus menziesii* is the sole tall tree, but there is usually more or less

N. cliffortioides. In the western part of the district, there is much dicotylous–podocarp forest, but in the eastern part pure *Nothofagus* forest rules, though, in some localities, podocarps are not altogether absent. From Lake Te Anau southwards, *N. menziesii* is dominant, though generally mixed with more or less *N. cliffortioides.* Certain species, commonly subalpine, occur in the undergrowth, particularly: *Hoheria glabrata, Pseudopanax lineare, Archeria traversii* var. *australis, Phyllocladus alpinus* and *Coprosma ciliata,* but the most important species are *Coprosma foetidissima, Wintera colorata, Nothopanax anomalum, N. simplex,* many hybrids of the swarm *N. simpanomalum, Coprosma colensoi, C. astoni* and hybrids between the last two and *C. foetidissima.* On the floor is a deep covering of bryophytes, mats of *Nertera dichondraefolia, Enargea parviflora* and *Libertia pulchella.*

When *Nothofagus fusca* is present, it is rather as colonies than as the dominant of a subassociation. But so much of the Fiord district is unknown that statements based on a few localities are most likely misleading.

Nothofagus menziesii *forest of the South Otago district.* This association is similar to the *N. menziesii* association of the eastern Fiord district of which it is a continuation, but it lacks the true high-mountain element, and it contains fewer mats, etc., of liverworts. In places, there is a small amount of *N. cliffortioides.*

The association is mostly confined to the western part of the district, and originally extended from the Longwood Range – where, to some extent, it is generally mixed with podocarps – to its junction with the Fiord forest-covering.

A number of areas of *N. menziesii* forest occur near Dunedin which have recently been studied intensively by J. Simpson and J. S. Thomson [13]. One on the Silver Peaks is a good many square kilometres in area. On the whole, the species are much the same as those of dicotylous–podocarp forest, but podocarps are wanting. The undergrowth also is far more open. Simpson and Thomson consider the areas as relics of a primitive *Nothofagus* forest and they supply strong evidence supporting this view.

Subantarctic–subtropical lowland-forest. Forest in which other dicotylous trees are present as well as one or more species of *Nothofagus,* together with podocarps, are common throughout the range of

Nothofagus. They are most abundant in the north and west of South Island, together with the South Otago district. Either the dicotylous–podocarp element may dominate or *Nothofagus*.

Ecologically, such mixed forests are more hygrophytic than pure *Nothofagus* forest and, when the trees of the two classes are in about equal proportion, there is little to distinguish the community from the ordinary subtropical forest of the locality. *Weinmannia racemosa* is generally abundant and may dominate just as in so many dicotylous–podocarp associations.

As to the origin of these mixed forests, speculation comes into play. It is well known that, although the species of *Nothofagus* can grow under more unfavourable conditions (poor soil, moderate rainfall, heavy wind, 'sour' ground) than the *tall* podocarps, yet they thrive best with good soil and a moist climate; in fact, there is nothing, so far as their 'likes' and 'dislikes' go, to hinder them from always growing in subtropical forest. But where the latter is fully established, there is not light enough in its interior for any species of *Nothofagus* to gain a footing. Once there, however, and if sufficient light is let into the forest, the *Nothofagus* is better able to take advantage of the situation than any other tall tree, thanks to its comparatively rapid growth and its light-demanding nature. In mixed forest in the North-western district, when a podocarp falls, seedlings of one or other of the species of *Nothofagus* generally take possession of the ground. Certainly, if shade-tolerating saplings already occupy the soil, the *Nothofagus* can do little, so that progress towards replacement will be slow enough. Nevertheless, it is not unreasonable to assume that in mixed forest generally *Nothofagus* has a good chance of being the climax. Certain forests show such change in progress, e.g. almost pure *Nothofagus* forest near Lake Te Anau with a few trees of *Dacrydium cupressinum*, or the miserable 'suppressed' saplings of *Podocarpus hallii* in so many *Nothofagus* forests near Lake Wakatipu. Indeed such replacement by *Nothofagus* is no uncommon thing. This phenomenon depends upon the rapid growth of a light-demanding tree, just as the *Beilschmiedia tawa* succession depends upon the shade-tolerating habit in the presence of slow-growing, light-demanding podocarps. The same result– eventual dominance – is thus attained by species of opposite ecological properties and requirements!

REFERENCES

[1] COCKAYNE, L. (1926) 'Monograph on the New Zealand beech forests: Part 1. The ecology of forests and the taxonomy of the beeches', *Bull. No. 4, N.Z. State Forest Service.*

[2] CHEESEMAN, T. F. (1897) 'On the flora of the North Cape District', *Trans. Proc. N.Z. Inst.*, XXIX 333.

[3] —— (1872) 'On the botany of the Titirangi District of the Province of Auckland', ibid., IV 270.

[4] KIRK, T. (1870). 'On the botany of the Thames goldfields', ibid., II 89.

[5] ADAMS, J. (1884) 'On the botany of the Thames goldfields', ibid., XVI 385.

[6] —— (1889) 'On the botany of Te Moehau Mountain, Cape Colville', ibid., XXI 32.

[7] WILSON, E. H. (1922) 'Notes from Australasia: II. The New Zealand forests', *Journ. Arn. Arboret.*, II 282.

[8] YEATES, J. S. (1924) 'The root-nodules of New Zealand pines', *Journ. Sc. & Tech.*, VII 121.

[9] CAMPBELL-WALKER, J. (1877) 'Report of the Conservator of State Forests', *Journ. House of Repr.*, C.3.

[10] SPEIGHT, R. (1911) 'The post-glacial climate of Canterbury', *Trans. Proc. N.Z. Inst.*, XLIII 408.

[11] ARMSTRONG, J. B. (1880) 'A short sketch of the flora of Canterbury, with catalogue of species', ibid., XII 324.

[12] ALLAN, H. H. (1926) 'Vegetation of Mount Peel, Canterbury, N.Z.: Part I. The forests and shrublands', ibid., LVI 37.

[13] SIMPSON, G., and THOMSON, J. S. (1926) 'Results of a brief botanical excursion to Rough Peaks Range', *N.Z. Journ. Sc. & Tech.*, VIII 372.

8 Coniferous Forests of North America

J. E. WEAVER and F. E. CLEMENTS

From *Plant Ecology*, 2nd ed. (McGraw-Hill, 1938) pp. 488–92, 496–504.

THE BOREAL FOREST

THE boreal forest of North America has itself been differentiated from an earlier circumpolar mass and, hence, stands in close relationship to the coniferous forests of northern Europe and Siberia. It is even more closely related to the subalpine forests of the Rocky Mountains and the Sierran-Cascade system, which are to be regarded as more recent climatic modifications of it. The elevation of the Appalachians has been too slight to produce such a result, though the presence of *Picea mariana* and *Abies fraseri* in the southern ranges is evidence of such a tendency, which is further confirmed by the zone of boreal forest on the high peaks of New England.

Because of the general uniformity of conditions and their very gradual change to the north and west, this climax is not so distinctly marked off into associations. The disappearance of balsam fir and jack pine, however, the increased importance of aspen and birch, and the entrance of alpine fir and lodgepole pine make the recognition of two associations necessary, though it must be admitted that the ecotone between them is a very broad one. The subclimax associes plays a much larger part than usual in most climaxes, owing to the fact that the subclimax dominants outnumber the climatic ones. Moreover, they not only take possession of fire-swept or lumbered areas but also assume a regular role in the succession found in the innumerable bogs and muskegs and on sandy or rocky plains [1].

The spruce–larch forest

This association reaches from Labrador, Newfoundland and New Brunswick on the east to the Rocky Mountains of northern British Columbia and the Yukon and in its extent across the continent is exceeded only by the arctic tundra [2]. In its strictest sense, its climax dominants are restricted to two species, the white spruce, *Picea glauca*, and the balsam fir, *Abies balsamea*, but the several subclimax trees may assume climax roles as well. This is primarily an outcome of the climatic relations as the tree limit is approached, in either latitude or altitude. Increasingly rigorous conditions cause the climax species to dwindle in importance or drop out, while the less exacting ones persist. Thus, the larch or tamarack, *Larix laricina*, which is typical of bogs or muskegs through the heart of the forest, becomes essentially a climax tree along the northern border. This is likewise true of the paper birch, *Betula alba papyrifera*, which is elsewhere characteristic of burns or of immature soils. The black spruce, *Picea mariana*, grows commonly with larch in or about bogs, but it becomes climax on rocky plateaux or on high mountain slopes, and the aspen, *Populus tremuloides*, often exhibits the same tendency. The jack pine, *Pinus banksiana*, appears to assume subclimax or climax roles with equal readiness, but its definitely subclimax nature in the more temperate lake forest indicates that it is usually climax in the boreal one. The large-toothed aspen, *Populus grandidentata*, belongs in lowland or fire subclimaxes and is rarely if ever a climax species. Two other trees are frequent in the eastern portion of this association, viz., arborvitae, *Thuja occidentalis*, and red maple, *Acer rubrum*. Both are more or less subclimax in nature, but they also persist well into the climax community. The former, however, is regarded as belonging properly to the lake forest and the latter probably to the deciduous one.

The characteristic undergrowth of the spruce–larch forest is supplied by the heath stage of the bog succession. The most important species are *Kalmia glauca*, *K. angustifolia*, *Ledum palustre*, *L. groenlandicum*, *Chamaedaphne calyculata*, *Rhododendron canadense*, *Empetrum nigrum*, *Rubus chamaemorus*, *Andromeda polifolia*, and several species of *Vaccinium: pennsylvanicum*, *caespitosum*, *uliginosum*, *vitis-idaea*, *oxycoccus*, etc. Most of these grow taller and more open as the larch and spruce close in on the moor, and the least tolerant species drop out as the canopy thickens with the entrance of balsam and white spruce. The margins of the moor are occupied by taller species, such as *Alnus incana* or *A. crispa*, *Viburnum cassinoides*, *V. pauciflorum*, *Corylus rostrata*, *Cornus stolonifera*, *Pyrus arbutifolia*, *Myrica gale*, *Betula pumila*, *Spiraea salicifolia*, etc., and of these *Alnus*, *Corylus* and *Spiraea* persist well into the shade. The most successful shade plants are the dwarf or creeping shrubs, such as *Gaultheria procumbens*, *Vaccinium oxycoccus*, *Cornus canadensis*, *Mitchella repens* and *Epigaea repens*, with which are associated

Coptis trifolia, Clintonia borealis, Pyrola elliptica, Chiogenes hispidula, Aralia nudicaulis, etc. The various species of *Sphagnum* usually disappear before the subclimax is reached, but *Polytrichum* and *Cladonia* sometimes persist into the climax.

The spruce–pine forest

This association covers northern British Columbia, the Yukon and Alaska up to the limits of the tundra in the north and to an altitude of 1000 to 2000 ft in the mountains. It possesses the white spruce in common with the eastern association, but the balsam fir and jack pine have disappeared, the larch is rare in Alaska, and the black spruce much less frequent. The three deciduous species, especially the paper birch, play a much larger part, and two new dominants enter from the subalpine forest of the Rocky Mountains, viz., lodgepole pine, *Pinus contorta murrayana,* and alpine fir, *Abies lasiocarpa.* In the north, even the white spruce becomes subordinated to the poplars and birches, though it persists in the south to Cook Inlet and beyond, where it is mixed with *Picea sitchensis, Thuja plicata* and *Tsuga heterophylla* of the coast forest. Owing to the fact that the peninsula of Alaska is largely surrounded by cold waters, and as well to the number of mountain ranges, its climate is arctic and tundra is the prevailing climax. It is chiefly along the Yukon river and its tributaries, and the Pacific Ocean, that forest climaxes are possible.

The bogs of black spruce and occasional larch exhibit many of the heaths and other shrubs of the eastern association, and the undergrowth of the climax areas is likewise much the same, until the influence of the Pacific Ocean is felt. In such regions, *Alnus, Cornus, Ledum, Ribes, Vaccinium* and *Viburnum* are joined by *Gaultheria shallon, Menziesia, Echinopanax,* etc., from the coast forest, and this undergrowth becomes controlling in mixtures of the two climaxes.

The birch–aspen associes

This is the characteristic fire subclimax throughout both associations. It is composed chiefly of the paper birch and aspen, though the balsam poplar and jack pine take some part in it. The first two may appear as pure or nearly pure consocies, or they may be mixed in various degrees, often with a sprinkling of balsam poplar. The latter is rarely pure, except occasionally on flood plains, but the jack pine usually constitutes a consocies, owing to the relation between its closed cones and seeding. Birch and aspen occur more or less abundantly as relicts in the climax forest, and particularly at the margins, where fire and clearing have been at work [3]. They are, however, at an increasing disadvantage in competition with the climax dominants as the latter grow taller, and they gradually drop out and finally disappear in the mature forest. The undergrowth is better developed than in the climax as a result of the more open canopy. It is often dominated by *Pteris aquilina*, which finally yields to the original forbs of the climax.[1]

THE LAKE FOREST

Extent and nature

As the name implies, this is pre-eminently a lake formation, being centred on the Great Lakes and recurring to the eastward in New York and New England where the larger lakes and rivers produced similar conditions. Since sandy soils likewise furnish favourable water and temperature relations, the pines, in particular, are to be found on sandy plains through much of this region. The most extensive stands of white pine, *Pinus strobus*, were originally found in central and northern Michigan and in eastern central and northern Minnesota. Farther east, the forest was more fragmentary, consisting chiefly of relict communities of varying size, found about bodies of water, on sand plains, or on the slopes of the Allegheny Mountains in Pennsylvania and to the southward. The climax dominants, white and red pine and hemlock, occur over a much wider area, smaller relicts, as a rule, persisting through southern Ontario and Quebec, much of New Brunswick and central Maine.

The climate of this forest has a wide geographic as well as annual range. The annual precipitation varies from a mean of 25 in. in Minnesota to one of nearly 45 in. in the mountains of the east. The temperature extremes during the year may range from −50 to 105°F., and in the northern portion frost may occur in any month of the summer. The growing season averages four months, though white pine and hemlock persist under favourable local conditions in regions where it is much longer.

[1] The student may well consult the following general sources of information on plant communities of North America for further details: Clements, 'Climax Formations of Western North America', in *Plant Indicators*; Shantz and Zon, *Natural Vegetation*; Shelford *et. al.*, *Naturalist's Guide to the Americas*; Harshberger, *Phytogeographic Survey of North America*.

The pine–hemlock forest

The lake forest consists of a single association, in which *Pinus strobus*, *P. resinosa* and *Tsuga canadensis* are the climax dominants. It has been so long cut off from the related montane and coast forests of the west that they cannot be grouped in the same climax, though their phylogenetic relationship seems evident. In fact, the climax nature of the lake forest itself may be easily questioned today, because of the great vicissitudes it has experienced. No other association has suffered so severely from lumbering and consequent fire, partly because of the quality of its timber, but chiefly because of its proximity to long-settled and well-populated districts. In southern Ontario where white pine with considerable red pine constituted formerly 60 per cent of the forest, logging and fire have reduced this to scattered relicts, about which effective reproduction is still further handicapped by the coactions of man. It is such universal disturbance that is primarily responsible for the doubts as to the actual existence of a pine–hemlock climax, but earlier historical and physical factors have had a large share as well [4, 5, 6].

During the repeated mass migrations of the glacial-interglacial cycles, this entire climax suffered not only the most severe buffeting but also the most intense competition from the boreal forest along one border and the deciduous forest on the other. Its migration before the ice front or in the wake of its retreat was, moreover, peculiarly handicapped by the solidarity of the great mass of the hardwood forest and, during the recent period, by that of the boreal forest as well, to say nothing of the barriers constituted by the Great Lakes. In a region with such marked extremes of climate, each phase of every major climatic cycle has increased its disadvantage. The cold-dry phases have permitted the encroachment of the boreal climax, the warm-wet ones that of deciduous forest, not over a uniform terrain but one fragmented by lakes, rivers and mountain ranges to the extreme. One striking consequence has been the inclusion of many small, relict areas of pine, hemlock or both well within the mass of deciduous or boreal forest. When all the evidence is assembled and weighed in the light of these various processes, there seems little doubt the pine–hemlock forest represents a genuine climax, now sadly depleted and fragmented, especially by the hand of man.[1]

[1] The most conclusive testimony has been furnished by Sargent, who saw this forest in much its original condition and set it apart as a distinct community [7].

The climax dominants of this association are red or Norway pine, *Pinus resinosa*, white pine, *P. strobus*, and hemlock, *Tsuga canadensis*. A frequent associate of the pines is the jack pine, *P. banksiana*, but this belongs properly to the subclimax. On mountain slopes the red spruce, *Picea rubra*, is associated with white pine especially but is, perhaps, even more frequent in montane communities of white spruce and balsam fir. Southward, *Pinus rigida* also enters this community and serves to connect it with the subclimax pine forest of the south-east. The arborvitae, *Thuja occidentalis*, often plays a role of considerable importance, but it, too, is to be regarded as a subclimax species. The white cedar, *Chamaecyparis thyoides*, likewise exhibits affinities with this group but is rarely of much importance even in the subclimax (Plate 8).

The difference in the ecological requirements of the three climax dominants is such that they are not frequent in mixture, but this has undoubtedly been, in part, an outcome of lumbering. The two pines occur together throughout most of the western portion, but the greater tolerance of the white pine for shade originally produced extensive pure stands [8, 9]. Over the eastern part of the area, the relict areas appear frequently to be either pure pine or hemlock, but in the original forest the two consociations grew side by side as well as in mixture [10, 11]. The view that hemlock is properly a member of the deciduous forest runs counter to the rule as to the identity of life-forms and has not taken sufficient account of the nature of relicts. Since its tolerance of shade and its water requirements are greater than those of the white pine, the hemlock approaches beech, maple and chestnut closely in its demands. Its proper climax position is disclosed, however, by the relict communities in the maple–beech association, where it is found all but invariably on the cool northerly slopes [12].

The number of genera common to the pine–hemlock association and the coast and montane forest of the west is so great as to indicate that they were originally derived from the same coniferous climax. The transition association of the coast climax, in particular, has a species corresponding to practically every one of the lake forest. *Pinus strobus* is represented by another white pine, *P. monticola*; the red pine by the ponderosa pine, *P. ponderosa*; and the jack pine by another species of the same character, *P. contorta*. *Tsuga canadensis* has a reciprocal species in *T. heterophylla*, *Thuja occidentalis* in *T. plicata*, and *Larix laricina* in *L. occidentalis*. The presence of

Chamaecyparis on both coasts is a further point of resemblance. The montane forest is more nearly related to the coast climax, but it also contains the three types of pine as well as a group of more recently evolved southern pines corresponding, in some measure, to the numerous species of the pine subclimax of the south-east.

The characteristic subclimax of the pine–hemlock forest is formed by the consocies of jack pine, *Pinus banksiana*. As in practically all species of this group, the cones not only remain on the trees for a number of years, but also they open tardily and sometimes only as a result of fire. This species is, in consequence, especially fitted to take possession of burned areas as pure stands; its preference for sandy plains is likewise to be explained by its lower requirements, though in such situations it may be the subfinal stage of the normal prisere. Birch and aspen also occur in a fire subclimax in this forest, though largely as a result of mixture with the boreal formation [13]. *Thuja occidentalis* is regarded as the typical subclimax of the hydrosere, usually forming a fairly distinct and nearly pure zone about the drier margins of bogs. At its inner edge it is frequently associated with *Picea mariana* and *Larix laricina*; though the latter belong properly to the boreal climax the frequent occurrence of relict bogs in the northern portions especially of the pine–hemlock climax gives these two species more or less subclimax importance in the latter.

The undergrowth of this forest is rather poorly developed, owing to the dense canopy, particularly in the hemlock consociation. Moreover, from its position, the species of shrubs and herbs are common to boreal or deciduous forest for the most part. They are necessarily shade plants of a more or less extreme type, largely ferns and fernworts, *Asplenium*, *Polystichum*, *Lycopodium*; orchids, *Calypso*, *Goodyera;* saprophytes, *Corallorrhiza*, *Monotropa*, *Hypopitys*; and such undershrubs and forbs of the ground cover as *Mitchella*, *Chimaphila*, *Gaultheria*, *Pyrola*, *Circaea*, *Viola*, etc.

THE COAST FOREST

Extent and nature

This climax has its greatest development along the Pacific Coast, as its name implies. The main body stretches from southern British Columbia to northern California, but several of the major dominants extend much farther northward as well as southward. *Picea sitchensis* finds its northernmost limit at Cook Inlet in Alaska, while *Tsuga heterophylla* and *Chamaecyparis nootkatensis* reach nearly as far. *Thuja plicata* occurs in southern Alaska, and *Abies amabilis* is found at the extreme southern end. *Sequoia sempervirens* ranges farthest to the south, its last outposts lingering in the Santa Cruz and Santa Lucia Mountains of California. While the best expression of this climax is along the coast, it extends to the Cascades and covers their western slopes. Eastward it passes into a broad transition forest that reaches to the western ranges of the Rocky Mountains in northern Montana and adjacent British Columbia. In altitude, the coast forest extends from sea-level to 3000 to 5000 ft in the coast ranges and the Cascades.

Geographically, this forest belongs to the coast and the Columbia Basin. At the higher levels, the latter resembles the former in being a region of relatively high rainfall and low evaporation. The exceptional extension along the coast is partly a matter of high rainfall but is due chiefly to the remarkable oceanic compensation between Alaska and California. The temperatures as well as the rainfall are less uniform from east to west, but this is reflected in the mixing of two climaxes and the differentiation of a transition community.

The closest relationship of the coast forest is with the montane climax, due, in some degree, to their direct contact at present. This is naturally best exemplified in the transition association, while the generic composition of the coast association is more like that of the pine–hemlock forest of the north-east, as already indicated. Both the former show an affinity with the boreal forest in the presence of *Abies* and *Picea*, though this may really be through the subalpine forest. The most important dominant in common is the Douglas fir, *Pseudotsuga taxifolia*, which reflects the climatic relations in being climax in the montane formation and subclimax in the higher rainfall of the north-west. The chief contact of the coast forest is with the montane, until this yields in the north to the subalpine and the latter to the western association of the boreal climax. About Cook Inlet the grouping may include a single dominant of each of the last three.

The cedar–hemlock forest

This is much the more massive and continuous of the two associations. The dominants are fewer and the composition less varied, though the northern and southern extremes show striking differences from the central portion. The trees are taller, the canopy denser, and the shrubby layer often developed to form almost impenetrable thickets.

The most typical expression is found in western Washington and British Columbia, whence the forest decreases in width and number of dominants in both directions, *Picea* and *Tsuga* forming the northern and *Sequoia* the southern outposts [14, 15].

The essential character of the coast forest is given by *Tsuga heterophylla*, *Thuja plicata*, *Picea sitchensis* and *Sequoia sempervirens*, though the typical grouping comprises the first two together with *Pseudotsuga taxifolia*. *Picea* is confined to the vicinity of the sea coast and *Sequoia*, though more southern, restricted chiefly to the fog belt; the several species of *Abies* and *Chamaecyparis* range well into the adjacent communities [16]. With respect to abundance and extent, *Pseudotsuga* is much the most important dominant. It is the typical species of burned areas and, hence, properly constitutes a subclimax, a conclusion further supported by its relatively low tolerance [17]. Much of the area formerly covered with *Tsuga*, *Thuja* and various associates is now a pure stand of Douglas fir, in which the hemlock and cedar persist in deep canyons or other protected places. This species is likewise a major dominant of the montane forest as well as of the transition association and, in consequence, passes readily from the role of subclimax to climax dominant.

This association constitutes a coniferous forest of unrivalled magnificence, in which the mature trees reach heights of 200 to 300 ft and diameters of 15 to 20 ft [18]. This is a direct outcome of the moderate temperatures and excessive rainfall with frequent or constant fog. Over much of the area the annual rainfall is in excess of 80 in., the range being 50 to 120 in. In the United States, all but 10 to 30 per cent of this falls during the six winter months, and much the same conditions obtain to Sitka and beyond. The absolute minimum as far north as Sitka is but $-4°F$. (Plate 9).

In a region of such excessive precipitation, the water relations of the dominants are less clear or at least are much modified by temperature. The general relation to the factor complex is indicated by the altitudinal range, though this is not in full accord with that of latitude. The typical fog-belt trees are *Picea sitchensis*, *Sequoia sempervirens* and *Chamaecyparis lawsoniana*, which indicate maximum conditions as to water content and humidity. These are followed by *Thuja plicata* and this by *Tsuga heterophylla* and *Abies grandis*. The ability of *Abies amabilis*, *A. nobilis* and *Chamaecyparis nootkatensis* to endure more xeric conditions is shown by the fact that they occur also in the subalpine zone, where the first is frequent at timber line. *Pseudotsuga* is the most xeric of all the dominants, a fact in complete accord with its subclimax nature and its importance in the montane forest.

Shrubby societies are characteristic of this forest and those of forbs are correspondingly reduced. *Gaultheria shallon* and *Echinopanax horridum* are two of the most typical subdominants, while *Acer, Berberis, Ribes, Rubus, Rhododendron, Sambucus* and *Vaccinium* are the genera represented by two or more species. Among the ferns, *Blechnum spicant, Polystichum munitum* and *Pteris aquilina* are widespread, while the ground cover is composed chiefly of *Oxalis oregana, Asarum caudatum, Fragaria vesca, Trientalis latifolia, Trillium ovatum, Disporum smithii, Streptopus roseus*, etc.

The larch–pine forest

This association occupies the eastern slopes of the Cascades below the subalpine zone. It stretches across the mountains of northern Washington into Idaho and north-western Montana, reaching its limit on the western slopes of the Continental Divide. It is found on the Gold and Selkirk ranges of British Columbia, the southern Bitterroot Mountains of Idaho, and the Blue and Wallowa ranges to the west [19, 20].

This is primarily a transition forest between the coast and the montane climaxes, but it occupies such a large area that it cannot well be regarded merely as an ecotone. Over most of the region the dominants of the coast forest are characteristic, and for this reason it is assigned to this climax. The trees, however, are reduced in size and the association is less dense and exclusive, owing to increasing remoteness from the coast. Of the four major dominants of the coast association, *Picea sitchensis* has disappeared, *Tsuga* and *Thuja* diminish in importance and then disappear, and *Pseudotsuga* shares the control with several other important species. Over most of this forest the rainfall is only 20 to 35 in., and 30 to 60 per cent of this falls between 1 April and the end of September, a proportion twice as great as in the cedar–hemlock forest. The mean temperature is about 7°F lower and the minimum ranges from −25 to −49°F.

The chief contact of the transition forest is with the montane climax, and, in consequence, its dominants are almost equally divided between the two formations. Five are derived from the coastal association and three from the montane forest, while *Pseudotsuga* belongs to all three but is here more of the montane type. *Pinus monticola* and

Larix occidentalis reach their optimum development in northern Idaho and the adjacent regions and may well be regarded as the two most typical dominants of this forest. Likewise, while *Abies grandis* ranges from the coast to north-western Wyoming, it is more characteristic of the transition region. Toward the east the major dominants of the coast association are the first to drop out, followed by those of the transition community, while *Pseudotsuga, Pinus ponderosa, P. contorta* and *Picea engelmanni* continue into the Rocky Mountains as chief dominants. The undergrowth varies in harmony with the behaviour of the dominants; it is essentially the same as in the cedar–hemlock forest in the western portion, becomes a mixture in the central, and finally passes over more or less completely into that of the montane forest in the east.

REFERENCES

[1] HARSHBERGER, J. W. (1911) *Phytogeographic Survey of North America* (G. E. Stechert and Company, New York).

[2] HALLIDAY, W. E. D. (1937) *A Forest Classification of Canada*, Dominion of Canada Dept. of Mines and Resources, Forest Service, Bulletin 89.

[3] MOSS, E. H. (1932) 'The vegetation of Alberta: IV. The poplar association and related vegetation of central Alberta', *J. Ecol.* xx 380–415.

[4] BROMLEY, S. W. (1935) 'The original forest types of southern New England', *Ecol. Mon.*, v 61–89.

[5] LUTZ, H. J. (1930) 'The vegetation of Heart's Content, a virgin forest in northwestern Pennsylvania', *Ecology*, xi 1–29.

[6] NICHOLS, G. E. (1935) 'The hemlock–white pine–northern hardwood region of eastern North America', *Ecology*, xvi 403–22.

[7] SARGENT, C. S. (1884) 'Report on the forests of North America, exclusive of Mexico', Tenth Census.

[8] BERGMAN, H. F., and STALLARD, H. (1916). 'The development of climax formations in northern Minnesota', *Minn. Bot. Studies*, iv 333–78.

[9] COOPER, W. S. (1913) 'The climax forest of Isle Royale, Lake Superior, and its development, *Bot. Gazette*, lv 1–44, 115–40, 189–235.

[10] Bray, W. L. (1915) 'The development of the vegetation of New York State', *N.Y. State Coll. For., Tech. Pub. 3*.

[11] NICHOLS, G. E. (1918). 'The vegetation of northern Cape Breton Island, Nova Scotia', *Trans. Conn. Acad. Arts and Sci.*, xxii 249–467.

[12] WHITFORD, H. N. (1901) 'The genetic development of the forests of northern Michigan: a study in physiographic ecology', *Bot. Gazette*, xxxi 289–325.

[13] KITTREDGE, J., and GEVORKIANTZ, S. R. (1929) 'Forest possibilities of aspen lands in the Lake States', *Univ. Minn. Agric. Exp. Sta., Tech. Bull.*, 60

[14] MUNGER, T. T. (1917) 'Western yellow pine in Oregon', *U.S. Dept. Agric. Bull.*, 418

[15] PIPER, C. V. (1906) 'Flora of the State of Washington', *Contr. U.S. Nat. Herb.*, xi.

9 The Broad-Sclerophyll Vegetation of California

W. S. COOPER

From 'The broad-sclerophyll vegetation of California', *Carnegie Institute of Washington*, publication No. 319 (1922) 20–7.

THE BROAD-SCLEROPHYLL COMMUNITIES

THERE are two Californian formations in which the broad-sclerophylls are the dominating element – the broad-sclerophyll forest formation and the chaparral formation. Each is in part climax, in part successional. Further, there is a broad-sclerophyll element of minor importance in the redwood forest, making a rather large part of its undergrowth. The grounds upon which the formations have been distinguished, and their range, composition and structure, will be given here.

The broad-sclerophyll forest formation is dominated by trees, mainly sclerophyllous evergreens, but including a number of deciduous species (30·8 per cent). It is typically climax, but in this phase its extent is limited. It is successional where its range overlaps the ranges of the conifer formations, and post-climax in its overlap with the climax chaparral.

The chaparral formation is made up of shrubs, the great majority being sclerophyllous evergreens. Its climax and successional phases are both of great importance. The latter is related developmentally to the conifer climaxes and is almost totally distinct floristically from the climax phase.

The broad-sclerophyll forest formation

This formation ranges from southern Oregon southward through the coast mountains and Sierra foothills into Lower California, reaching its limit probably in the region of Mt San Pedro Martir. Nowhere, so far as I am aware, does it dominate the country as a conifer forest, for instance, commonly does. It occurs rather in discontinuous patches, which may, however, be of considerable extent. These alternate in the main with patches of chaparral of the type which I have designated as climax. Northward the forest is the more important of the two, especially in the Coast Ranges, while southward the chaparral becomes more and more preponderant. In the north there is overlap also with the ranges of the *Sequoia sempervirens* and the *Pseudotsuga* associations of the Pacific conifer formation, and in the Sierras with the formations of the conifer forest region.

The number of dominant species is not large. In the following list a single asterisk indicates importance also in the conifer forest chaparral; and two asterisks, in both that and the climax chaparral.

Sclerophylls

Myrica californica	*Quercus chrysolepis*	*Pasania densiflora**
Castanopsis	*engelmanni*	*Umbellularia*
chrysophylla*	*wislizeni***	*californica*
Quercus agrifolia		*Arbutus menziesii*

Deciduous

Quercus kelloggii*	*Acer macrophyllum*	*Aesculus californica*
lobata		

Several associations occur, easily recognised because of their relative constancy and wide distribution. Many individual localities would fit into none of them, and therefore in a minute study numerous transitional units might be described.

Pasania–Quercus–Arbutus association. This community is characteristic of the lower altitudes of the North Coast Ranges, a region very complex and rather difficult to understand vegetationally, since two or more types which are climactic nearby meet here and overlap, finding conditions that are reasonably favourable to all. The redwoods thoroughly dominate the coast. East of them *Pseudotsuga mucronata* is the commanding species, but shares its rule with the broad-sclerophyll association about to be described. Chaparral and grassland communities also occur, but these are plainly successional. The *Pasania–Quercus–Arbutus* association is itself somewhat of a transitional unit between broad-sclerophyll and conifer types, for it rarely occurs without at least a sprinkling of conifers, especially *Pseudotsuga*, and its principal species occur commonly as an understory of the *Pseudotsuga* and *Sequoia* forests.

The most important species of the formation are *Pasania densiflora*, *Quercus chrysolepis* and *Arbutus menziesii*, and these attain great size. Other tree species occurring more or less commonly are *Quercus kelloggii*, *Castanopsis chrysophylla*, *Umbellularia californica*, *Acer macrophyllum, Æsculus californica* and *Cornus nuttallii.* The association ordinarily possesses two layer societies – one of shrubs, including *Corylus rostrata californica, Vaccinium ovatum* and *Gaultheria shallon,* and one of herbs and ground-shrubs, a mixture of typically oak-forest species and those commonly associated with the redwood and Douglas fir.

The transitional phases of the association will be made evident by description of two areas, one in the interior of the Coast Ranges, where *Pseudotsuga* is the competing tree, the other on the edge of the coastal redwood region.

The first locality is in Trinity County, on the north-facing slope of the Mad-Trinity Divide. The dominant tree, *Pseudotsuga,* grows here magnificently, many specimens attaining a diameter of 2 m. *Abies concolor* and *Pinus lambertiana* also occur. Beneath the conifers there is an understory of broadleaf trees, nearly all sclerophylls. *Castanopsis chrysophylla* is the most abundant, and *Pasania, Arbutus* and *Acer macrophyllum* also occur. This assemblage might here be termed a layer society. *Corylus rostrata californica, Cornus nutallii* and *Ceanothus integerrimus* form a second stratum, and a third is composed of herbs and ground shrubs: *Vancouveria* sp., *Polystichum munitum, Berberis* sp., *Gaultheria shallon.*

The other is the valley of the South Fork of the Eel river, in Humboldt County. In the vicinity of Garbersville and for several kilometres north of it (downstream), the north slopes and ravines are forested with *Pasania densiflora, Quercus kelloggii, Q. chrysolepis, Q. garryana, Arbutus menziesii, Umbellularia californica, Æsculus californica, Acer macrophyllum* and *Pseudotsuga,* making a rather typical specimen of the *Pasania–Quercus–Arbutus* association. The first redwoods appear in groups of large trees on the valley bottom, with scattered individuals on north slopes. This continues for 15 km. or more, then for several kilometres there is an almost pure forest of *Sequoia* on north slopes with the *Pasania–Quercus–Arbutus* association on south exposures. Finally the forest becomes nearly pure *Sequoia* on all slopes, the broad-sclerophylls gradually disappearing as a distinct community, though remaining to some extent as a layer society, particularly *Arbutus* and *Pasania.*

Quercus agrifolia–Arbutus association. This unit is coastal, occurring from the northern limit of *Quercus agrifolia* in Mendocino County to the southern limit of *Arbutus* in Los Angeles County. It is thus characteristic of the outer central Coast Ranges, where it is the dominant cover on north-facing slopes. It is particularly well developed in the San Francisco Bay region and southward to the Santa Lucia Mountains. The character tree is *Quercus agrifolia*.[1] *Arbutus menziesii* is next in importance, but varies greatly in abundance in different localities. *Æsculus californica*, a deciduous species, is usually prominent, and *Umbellularia californica* is equally so. *Acer macrophyllum* is frequently important in the more mesophytic localities. In areas that are transitional with the *Sequoia* association, *Pasania*, *Quercus chrysolepis*, *Q. kelloggii* and *Sequoia* itself occur. A typical area of this association occurs around Station 7 at Jasper Ridge (Plate 10(*b*)).

Quercus agrifolia *consociation*. South of the southern limit of *Arbutus* (Los Angeles County) the community is continued as a consociation dominated by *Quercus agrifolia*. This is rather prominent in the lower altitudes of the west slope of the Cuyamaca Mountains.

Umbellularia *consociation*. *Umbellularia* occurs scattered through the *Quercus agrifolia–Arbutus* association, and in others as well, but it also forms pure growths, especially in the central Coast Ranges, occupying moist ravines and canyon bottoms. These groups of *Umbellularia* stand out strikingly above the other trees, being conspicuous by reason of their light-green colour and conifer-like form. The shade is very dense and undergrowth almost lacking. *Umbellularia* itself, however, is able to germinate successfully under such conditions.

Quercus agrifolia–lobata *association*. This is characteristic of the broad valleys and gentle footslopes of the central Coast Ranges, being locally of considerable importance in the San Francisco Bay region. The dominant species are *Quercus agrifolia* and *Q. lobata*, the latter being deciduous. The trees as a rule stand far apart, producing a park-like landscape, and it is in such places that the largest specimens of both species occur. One tree of *Q. agrifolia* near Palo Alto is 2·1 m. in diameter breast-high. A specimen of *Q. lobata* west of Clear Lake is of the same diameter, with a spread of branches of 47 m. Much larger trees of the latter have been reported. Other

---

[1] In some places, especially in the inner Coast Ranges, it is replaced by *Q. wislizeni*, and thus another association might be distinguished.

species are of occasional occurrence. In the Palo Alto region large specimens of *Umbellularia*, *Arbutus* and *Prunus ilicifolia* make a small part of this association. Because of the wide scattering of the trees, the ground between is in most places under cultivation. Near Palo Alto, however, there are a few localities which retain their original vegetation, because they have long been included in certain large estates. In such areas one finds the two oaks of all sizes from seedlings to large mature trees. Most of the young ones occur in indefinite groups in the opener places, while the mature specimens completely dominate the ground beneath them. Three layer societies occur. The first, of tall shrubs, includes *Rhamnus californica*, *Heteromeles arbutifolia*, *Sambucus glauca* and *Rhus diversiloba*. The low-shrub society includes *Rubus vitifolius*, *Symphoricarpos racemosus* and *Solanum umbelliferum*. *Micromeria chamissonis* is dominant in the ground layer society. Further details concerning this very interesting association will be given in a future paper upon the communities and successions of the Palo Alto region.

Quercus chrysolepis–kelloggii association. The associations so far described are distinctly of low altitudes. The present one belongs to the higher Coast Ranges and southern California mountains and to the middle altitudes of the Sierras. It is pre-eminently a north-slope forest, but localities are common enough where it occurs on other exposures as well, seeming like a true climax. The most important tree species is the broad-sclerophyll *Quercus chrysolepis*; *Q. kelloggii*, deciduous, is often a close second. Others are *Arbutus*, *Umbellularia*, *Acer macrophyllum*, *Pasania*, *Æsculus*. Since the association has so great a range, the subordinate vegetation varies greatly. It shows broad transition areas with neighbouring associations. Its close relation to the *Pasania–Quercus–Arbutus* association is at once evident, and it is not strange, therefore, to find areas that cannot be placed with certainty in either. Again, just as the *Pasania–Quercus–Arbutus* association passes into the *Sequoia* forest as an understory, so also does the *Quercus chrysolepis–kelloggii* association into the pine forests of the high Coast Ranges and the Sierras. In fact, in the Sierras it is commoner to find the community as an understory beneath *Pinus ponderosa* and *P. lambertiana* than as a dominating type. In the mountains of southern California the group forms a similar understory beneath *Pseudotsuga macrocarpa*. Upon the xerophytic side there is transition to the chaparral. Such areas have so individual a stamp that I have been accustomed to refer to them as 'dwarf forest'. An excellent example is found upon the north slope of Mt Tamalpais (Marin County), near the summit. *Quercus chrysolepis* and *Q. wislizeni*, growing in dense thickets 3 to 5 m. in height, are dominant, an occasional full-sized tree of *Q. chrysolepis* rising above the general level. With them grow other species: *Quercus agrifolia, Pasania, Arbutus, Umbellularia, Torreya californica*, and the chaparral shrubs *Arctostaphylos tomentosa* [*A. glandulosa* Eastwood], *Ceanothus sorediatus* and *Castanopsis chrysophylla minor*.

Quercus chrysolepis *consociation*. In the middle altitudes of the Sierras, dominated by the pines and *Pseudotsuga, Quercus chrysolepis* growing almost pure has a distinct successional role. Upon the great talus accumulations at the bases of the Yosemite cliffs certain chaparral shrubs are the pioneers. These are followed by a dense, pure growth of *Quercus chrysolepis* which seems to persist for a long time, as the live-oak forest is the most conspicuous feature of such areas. The talus piles that are manifestly oldest, with much accumulation of humus, support a mixture of the oak and *Pseudotsuga*.

A few concluding remarks in summary will gather together the main points in the discussion of the broad-sclerophyll forest formation. It is plain that the group as a whole is the fundamental unit, the minor divisions being closely tied together by a number of binding species. The transition zones between associations of the formation and with other formations are broad, so that accurate delimitation is difficult. The broad-sclerophyll communities, wherever they adjoin the conifer forest communities, pass into them as layer societies. There is a very close habitat relation between the broad-sclerophyll forest and the climax chaparral, in that in the main they overspread the same range, occupying areas of comparatively slight physical differences. The question of climax, therefore, whether one or the other or both, is difficult.

The climax chaparral formation

The climax chaparral is the dominant community over the whole of the southern Coast Ranges and the mountains of southern California and northern Lower California. Only the highest summits, controlled by conifers, and the more mesophytic north slopes, inhabited by broad-sclerophyll forest, must be excepted. Northward in the north Coast Ranges the chaparral shares its control more and more with the broad-sclerophyll trees, and opposite the northern end of the Sacramento Valley it disappears entirely as a dominating community. In the southern Sierras it is of great importance, occupying a wide belt in the foothills. In the northern Sierras its continuity is broken, and this is not strange, since the conifers of the montane forest here reach the valley floor. The present range of the climax chaparral is indicated in a very general way by the range of its most important species, *Adenostoma fasciculatum*. In addition, I believe that there are certain extensive areas now inhabited by grasses and by half-shrubs that climatically and potentially are chaparral regions.

Since the climax chaparral is by far the most widely extended and diversified of the broad-sclerophyll communities, it is natural that the present list of species should be the longest. The following are all evergreen, except that *Quercus dumosa* is barely so. One asterisk indicates that the species is also of importance in the conifer forest chaparral; two asterisks that it is important in that and also in the broad-sclerophyll forest.

Castanopsis chrysophylla	*Rhus integrifolia*	*Garrya elliptica*
*minor**	*laurina*	*Comarostaphylis diversifolia*
*Quercus chrysolepis***	*ovata*	*Xylococcus bicolor*
dumosa	*Rhamnus californica**	*Arctostaphylos andersonii*
durata	*crocea*	*glauca*
wislizeni	*Ceanothus crassifolius*	*hookeri*
*frutescens***	*cuneatus*	*manzanita*
Dendromecon rigidum	*dentatus*	*montana*
Heteromeles arbutifolia	*divaricatus*	*pumila*
Cereocarpus betulaefolius	*hirsutus*	*stanfordiana*
Adenostoma fasciculatum	*megacarpus*	*tomentosa*
sparsifolium	*papillosus*	*vestita*
Prunus ilicifolia	*rigidus*	*Eriodictyon californicum*
Xylothermia montana	*sorediatus*	
Cneoridium dumosum	*verrucosus*	

With so large a list of species there is naturally great diversity in the composition of the association. Anyone given to splitting of hairs would easily separate many communities of lower rank. This is in part due to slight habitat differences, but also in an important degree to the great number of species with restricted range and to the frequent occurrence of fires, which result in multitudinous combinations of species depending upon which are able to survive or to repopulate the area burned. It is easy to recognise, however, throughout the length and breadth of the region, one striking and characteristic consociation, for *Adenostoma fasciculatum* covers many hundreds of square miles in practically pure dominance. Other species, too, completely control certain areas, but it is far commoner for these to mingle with each other and with *Adenostoma* in an endless number of combinations and proportions (Plate 10(*a*)). In 87 listed localities in all parts of the State the following occurrences of important and widespread species are noted:

Adenostoma fasciculatum	75	*Cercocarpus betulaefolius*	19
Arctostaphylos (*all species*)	50	*Quercus wislizeni frutescens*	11
Heteromeles arbutifolia	26	*Rhamnus californica*	10
Ceanothus cuneatus	25	*Quercus chrysolepis*	10
Quercus dumosa and			
Q. durata	23		

Such being the condition it avails little to attempt to distinguish minor units within the association. It is more reasonable to express the differences by noting the dominance of one or more species in particular cases. One fact, however, must be brought forward. The genus *Arctostaphylos* gives its stamp to certain localities in a very characteristic way. No one species is dominant throughout. *Arctostaphylos tomentosa* is by far the most important, ranging over the whole region. *A. glauca* is abundant in the southern half of the State and *A. manzanita* in the northern, and several others are prominent locally. This phase nearly everywhere accompanies the *Adenostoma* consociation, occupying the less xerophytic north-facing slopes where these are not sufficiently moist to permit the forest to exist, and at higher altitudes replacing the *Adenostoma* consociation on the south slopes, the north exposures being forested.

10 The Mountain and South-western Flora of South Africa

J. W. BEWS

From *Plant Forms and Their Evolution in South Africa* (Longmans, Green, London, 1925) pp. 151–61.

THE mountain and south-western vegetation, though dominated by sclerophyllous shrubs, is not composed entirely of this one class of growth forms. There is almost as great diversity as in the tropical–subtropical regions, though it has been found convenient to deal with the vegetation of the region as a whole, since other types of growth form are subordinate. There are no grassland areas, and the numerous grasses occur sparsely scattered in tufts among the dominant shrubs.

The true Macchia may be from 10 to 20 ft in height, or, in the case of some of its component species, even higher.

Below this, there is every gradation down to a low type of heath, with such species as *Blaeria ericoides* dominant. It is similar in many respects to the *Calluna* of Europe. Over large areas, especially where the Veld has been interfered with, the Rhenosterbosch (*Elytropappus rhinocerotis*) is completely dominant. It is also a heath-like shrub, though it belongs to the Compositae.

The heath types often form a stage in the plant succession to the climax Macchia, just as the latter often gives way to forest. All over the region there are subordinate types of growth form, water plants, marsh plants, geophytic plants, climbing plants, psammophilous and halophytic plants, annuals, etc.

In the account which follows, attention is directed first of all and chiefly to the dominant sclerophyllous shrubs.

Among the Ericaceae the huge genus *Erica* is the most important with about 480 African species, mostly south-western. A few species extend eastward and northward through the tropics, including *Erica arborea* on the mountains of East Central Africa. Twenty-three species are recorded for Natal, of which two or three descend to the coast-belt, but the majority are confined to the mountain ranges, where they are prominent. Other genera with eastern extensions are *Philippia*, *Blaeria* and *Ericinella*.

While the Ericaceae are common in Europe, Asia and North and

South America, they are almost absent from Australasia, where they are replaced by the nearly allied Epacridaceae. In tropical regions they occur at higher altitudes, but have there in *Erica arborea* retained the more ancient tree growth form, as also in the Asiatic Rhododendrons. The fact that the Australian Epacridaceae (which differ chiefly in the stamens) are ranked as a separate family may possibly indicate that the South African Ericaceae have been evolved since the connections with Australia – if these ever existed – were broken.

Among the South African Proteaceae one genus, *Faurea*, which reaches tree size, has 15 species and is purely subtropical and southeast African. It is prominent in the Brachystegia tree-veld of southern Rhodesia, and one species, *Faurea saligna*, is often completely dominant on sandy soil in one or two localities in Natal.

The genus *Protea* (130 species) is mostly south-western. Some of the Sugar-bushes, e.g. *Protea mellifera, P. grandiflora, P. lepidocarpodendron, P. neriifolia*, are among the commonest of the Macchia shrubs on mountain slopes. A few species extend eastward and northward through the tropical mountains to Abyssinia (*P. abyssinica*). Nine or ten species are recorded for Natal, several of them small trees which are dominant in the 'Protea Veld' of the Drakensberg and other mountains, growing isolated among the subtropical grasses. *P. roupelliae* is one of the commonest. *P. hirta*, a small shrub, descends to the coast.

Leucadendron has 75 species, all south-western, including the well-known *L. argenteum* (the Silver Leaf), which reaches a height of 50 ft, but according to Sim [1] 'is native only in the neighbourhood of Capetown'. *L. plumosum* is one of the common Macchia shrubs. *Leucospermum* has 40 species in the south-west, and there is one somewhat uncertain record, *L. gerrardi*, for the Natal Drakensberg. *Leucospermum conocarpum* is often dominant in the Macchia of the Cape Peninsula.

Other genera are *Serruria* (70 species), *Spatalla* (25 species), *Sorocephalus* (12 species), *Mimetes* (20 species), *Paranomus* (15 species), *Spatallopsis* (5 species), *Aulax* (3 species) and *Brabeium stellatifolium*, all south-western. The monotypic genus *Dilobeia* occurs in Madagascar, and with *Brabeium* makes up the only African representatives of the section Persoonieae.

The Compositae are also very common in the mountain and southwestern flora, but, though numerous, are, as a rule, subordinate ecologically to other Fynbosch shrubs. They more often belong to the heath stage of the plant succession in which *Elytropappus rhino-cerotis* is often dominant. Associated with it are other species belonging to the same family, e.g. *Metalasia muricata, Euryops tenuissimus, Eriocephalus umbellatus*. The most important genus of the Compositae is *Helichrysum*, of which there are about 30 species in the Cape Peninsula alone, many of them common. As large or even larger is the genus *Senecio*. Both *Senecio* and *Helichrysum* are widespread over the whole of South Africa, as well as *Berkheya, Athanasia, Osteospermum, Ursinia*, etc. More characteristic of the mountains and south-west are the genera *Felicia, Corymbium, Pteronia, Mairea, Helipterum, Metalasia, Stoebe, Pterothrix, Disparago, Anaxeton, Nestlera, Relhania, Athrixia, Printzia, Othonna, Euryops*. Many of these are common not only in the south-west but also along the Drakensberg.

Some of the common species even are thus widespread, e.g. *Metalasia muricata*. The majority of the mountain and south-western Compositae are more or less woody, though often much dwarfed. Some are woody only at the base, some entirely herbaceous, and a fair number are annuals.

A number of very characteristic though small south-western families may now be grouped together. The Bruniaceae have 12 genera and 55 species. Many of them are rather common in moist places and along stream-banks, e.g. *Brunia nodiflora, Berzelia lanuginosa, Berzelia abrotanoides*. *Staavia* is another important genus. *Berardia trigyna* is recorded for Natal. The Penaeaceae are a small family of heath-like shrubs and undershrubs, including 5 genera and 35 species (*Glischrocolla, Endonema, Penaea, Brachysiphon* and *Sarcocolla*). The Grubbiaceae include only one genus, *Grubbia*, with four species. In the Myoporaceae *Oftia jasminum* is common.

In the Rutaceae the south-western section of shrubs and undershrubs (the Rutoideae) with dehiscent fruits contrasts strongly with the trees of the eastern side. The most important western genera are *Agathosma, Barosma, Diosma, Phyllosma, Adenandra, Coleonema, Acmadenia, Macrostylis* and *Euchaetes*, including between them about 200 species, with only one, *Barosma lanceolata*, recorded for Natal. One genus, *Thamnosma* (with two species), has a curious distribution in Hereroland (South-West Protectorate) and in the island of Socotra. A detailed analysis of the whole family from the standpoint of origins and distribution would repay attention.

In the Geraniaceae the large genus *Pelargonium* (with 250 species) is a mountain and south-western type extending northwards across the tropics and producing a few species that are adapted as associated plants to grass-veld conditions, of which *P. aconitiphyllum* is abundant all through Natal. Several of the 17 recorded Natal species, however, are confined to the mountains, though nearly a dozen of them mix with the subtropical flora. The section Hoarea, consisting of stemless tuberous-rooted species, are all south-western. The section Peristera is most widely distributed, extending all over Africa, with one species in India and two in Australia. Floristically it is probably most primitive, and its habit is herbaceous like that of a Geranium. *P. grossularioides* belongs to it and extends from the Cape to the Drakensberg. The woody habit and the succulent habit seen in the Karroo species would seem to be derivative in this genus.

The Leguminosae (Papilionatae) are well represented by numerous species belonging to the genera *Cyclopia, Podalyria, Liparia, Priestleya, Amphithalea, Borbonia, Rafnia, Coelidium, Lebeckia, Viborgia, Lotononis, Aspalathus, Crotalaria, Argyrolobium, Psoralea, Indigofera, Tephrosia, Lessertia, Hallia, Virgilia, Rhynchosia* and others. As in the case of the Compositae, some of the larger genera are distributed all over South Africa. The relatively primitive Podalyrieae are more or less confined to the mountains and south-west. *Sutherlandia frutescens, Psoralea pinnata, Tephrosia capensis, T. grandiflora, Aspalathus laricifolia, A. canescens* and *A. spinosa* are species which extend from the Cape to Natal.

The Rosaceae are, on the whole, a temperate family in South Africa, though *Rubus pinnatus, Pygeum africanum* and species of *Parinarium* are subtropical. The most important mountain and south-western genus is *Cliffortia*. *C. linearifolia* is dominant in much of the Fynbosch of the Drakensberg, while *C. strobilifera* is widespread from the Cape Peninsula to Natal. One or two species of *Cliffortia* have invaded subtropical areas. *Leucosidea sericea* is an interesting monotype completely dominant in the Oudehout scrub of the Drakensberg, often forming a stage in the succession to high forest, as other Macchia shrubs (e.g. *Virgilia capensis*) do in the south-west. *Myrsine africana* (Myrsinaceae), which extends from the Cape to the Drakensberg, is another very important shrub. Like *Leucosidea*, it also forms a stage in the plant succession to scrub or forest; in fact, it is often ousted by *Leucosidea* itself.

In the Thymelaeaceae the heath-like genus *Passerina* has several species important in the south-western Fynbosch. Other species are scattered over South Africa. Other south-western heath-like genera of the Thymelaeaceae are *Cryptadenia* (5 species), *Lachnaea* (19 species), *Struthiola* (38 species, of which 2 reach Natal). The large genus *Gnidia* is widespread. *Myrica* spp. (Myricaceae) are common in moist places and near the seashore all over South Africa. *Osyris abyssinica* (*Colpoon compressum*), the Cape Sumach, is common at the Cape and extends over the eastern side of South Africa.

In the Rhamnaceae the genus *Phylica* has a large number of south-western and mountain species.

The Selaginaceae have the genus *Selago*, with 112 species in South Africa concentrated for the most part at the Cape, but 22 reach Natal, only one or two of which descend from the mountains. Two of the South African species reach the mountains of the tropics, and there are also 17 endemic species there. *Walafrida* has 31 South African, chiefly south-western species, 1 of which extends to the tropics, 4 in tropical Africa and 1 in Madagascar. All the species of *Dischisma* (11), *Microcodon* (5), *Agathelpis* (3), and *Gosela* (1) are south-western.

Other Dicotyledonous families may be grouped together and only the most important genera will be mentioned. Ranunculaceae (*Ranunculus, Knowltonia*), Cruciferae (*Heliophila*, 22 species in the Cape Peninsula), Polygalaceae (*Polygala, Muraltia, Mundia spinosa*), Caryophyllaceae (*Silene*), Sterculiaceae (*Hermannia*), Zygophyllaceae (*Zygophyllum*), Oxalidaceae (*Oxalis*, 32 species in Cape Peninsula), Crassulaceae (*Crassula, Rochea, Cotyledon*), Droseraceae (*Drosera*, 7 species in Cape Peninsula), Aizoaceae (*Mesembrianthemum, Tetragonia, Aizoon, Galenia, Pharnaceum, Adenogramma*), Umbelliferae (*Hydrocotyle, Anesorhiza, Peucedanum*; *P. galbanum* is very common on Table Mountain), Rubiaceae (*Oldenlandia, Anthospermum, Carpacoce, Galium*), Campanulaceae (*Lobelia, Laurentia, Cyphia, Lightfootia, Wahlenbergia, Microcodon, Roella, Prismatocarpus*), Plumbaginaceae (*Statice*, along seashore), Asclepiadaceae (various outliers), Gentianaceae (*Sebaea, Chironia Belmontia*), Borraginaceae (*Lobostemon*), Solanaceae (*Solanum, Lycium*), Scrophulariaceae (*Diascia, Nemesia, Zalusianskya, Phyllopodium, Sutera, Manulea, Melasma, Harveya*), Lentibulariaceae (*Utricularia*), Verbenaceae (*Campylostachys, Stilbe, Bouchea*), Labiatae (*Mentha, Salvia, Stachys, Leonotis leonurus*), Polygonaceae (*Polygonum*), Loranthaceae (*Viscum*), Santalaceae (*Thesium, Thesidium*), Euphorbiaceae (*Euphorbia, Cluytia*).

The above are, for the most part, subordinate types, often herbaceous. While some of them are more or less confined to the south-west, many of them are equally common as associated plants of eastern grass-veld areas. The widespread character of subordinate genera, and especially those that appear early in the plant succession, has been already referred to.

Among the Monocotyledons the most characteristic south-western family is the Restionaceae. They are practically confined to the south-west of South Africa, Australia, New Zealand and Tasmania, with one species in Natal (*Leptocarpus paniculatus*). The species of *Restio* formerly recorded for Natal was in error. There are also one species from Mlanje in South-East Africa (altitude 7000 ft), one species in Cochin-China, and one in Chile. The South African genera (about a dozen altogether, with 230 species) grow among the Macchia shrubs on dry hillsides or are often dominant in marshes and along streams. They are very xerophytic, with leaves usually reduced to a sheath.

The most important genera are *Restio* (100 species), *Elegia* (30 species), *Hypolaena* (20 species), *Thamnochortus*, *Hypodiscus*, *Dovea* and *Leptocarpus* (15 species each). The nearly allied Centrolepidaceae are Australian with representatives in South America, while the Eriocaulaceae, which are also possibly connected, are more tropical and subtropical.

The south-western grasses are, as a rule, entirely subordinate to the Macchia shrubs, though occasionally various species of *Danthonia* or *Pentaschistis* are dominant over small patches in early stages of the succession, while *Stenotaphrum glabrum* is sometimes dominant in moist places or on sandy flats.

The most important south-western genera of grasses are *Pentaschistis* (40 species), *Danthonia* (30 species), *Pentameris* (5 species), *Achneria* (9 species), *Ehrharta* (25 species), *Avenastrum* (7 species), *Lasiochloa* (3 species), *Brizopyrum* (5 species).

In the Liliaceae the larger genera are widespread and well represented in eastern grass-veld as in the south-west, e.g. *Anthericum*, *Urginea*, *Ornithogalum*, *Albuca*, *Chlorophytum*, *Bulbine*, *Asparagus*. *Lachenalia*, however, is characteristic of the south-west.

In the Amaryllidaceae the same thing is seen. *Hypoxis*, *Nerine* and *Cyrtanthus* are widespread, while most other genera are more eastern with outliers in the south-west.

In the Iridaceae, *Moraea*, *Aristea*, *Hesperantha*, *Watsonia* and *Gladiolus* are all large widespread genera, but there are a number of small south-western genera: *Ferraria, Hexaglottis, Galaxia, Witsenia, Babiana, Melasphaerula, Sparaxis, Synnotia. Romulea* and *Bobartia* are larger and have outlying species along the Drakensberg.

The Orchidaceae are very well represented in the south-west, but again nearly all the larger genera are widespread over South Africa: *Eulophia, Holothrix, Satyrium, Disa, Corycium, Disperis.* Other genera are wholly eastern and a few are characteristically south-western. Among the latter are *Orthopenthea, Amphigena, Schizodium, Evota, Anochilus.*

All the other smaller families of Monocotyledons belong entirely to subordinate (mostly marsh) forms. The subordinate associated plants of the whole region differ from those of the tropical–subtropical areas only to a slight degree. Aquatic and marsh types are very similar all over South Africa, and many of the species, as we have seen, are widespread. The Restionaceae, however, are more or less confined to the south-west. Climbing plants, undergrowth and the associated bulbous and tuberous plants and grasses differ from those of the eastern side only in a general increase of xerophytism, and among these the genera, as a rule, if not the species, are also widespread.

The dominant shrubs of the Macchia or Fynbosch are, however, of a very distinct ecological type well expressed by the term 'sclerophyllous', which has long been applied to it. They belong to Clements' class of Drymophytes. Their main evolutionary tendencies may be summed up briefly as follows:

1. Increased hardness of the leaves. Their leaves are full of fibre variously arranged, and though further experimental proof is desirable, there is little doubt that their resistance to water loss under conditions of extreme drought is thereby increased. Our experiments on various Natal species in this connection have already been referred to [2]. The comparative rates of water loss among the south-western shrubs have, however, not yet been much investigated. It is obvious that they do show extraordinary powers of resisting water loss during the intense heat of the long dry summer.

2. Reduction in the size of the leaves. This is seen in the majority of the species. The Proteaceae, however, and other taller trees or shrubs, e.g. *Olea* spp., have retained the flat type of leaf, relying on the increased hardness only. These taller forms are dominant in climax Macchia, and since succession here as elsewhere is towards the mesophytic they may be considered more mesophytic than the

lower-growing heath-like forms which appear earlier in the succession. Reduction in size of the leaves culminates in the next class.

3. The ericoid type of leaf. The leaf tends to curve downwards at the margins, enclosing the stomata on the under side. All stages of infolding may be seen. Heath-like forms are seen in a great variety of separate families. *Cliffortia* among the Rosaceae shows it, especially *C. linearifolia,* so common on the Drakensberg. The small families Penaeaceae, Geissolomataceae, Grubbiaceae and Bruniaceae are all more or less heath-like. The Penaeaceae are allied to the Thymelaeaceae, where *Passerina, Struthiola, Lachnaea,* etc., are also mostly ericoid.

Hieronymus [3] suggests that the Grubbiaceae represents the prototype of the Santalaceae, where *Thesium* and *Thesidium,* usually hemiparasites, continue the ericoid form.

In the larger families among the Leguminosae important genera such as *Amphithalea* and *Aspalathus* (150 species) are mostly heath-like, and among the Compositae species of *Stoebe, Disparago, Metalasia,* etc., are similar. Another heath-like composite is the common, widespread and often dominant Rhenosterbosch (*Elytropappus rhinocerotis*). The numerous Rutaceous shrubs of the southwest listed above are, to a large extent, ericoid. Among the Verbenaceae, species of *Stilbe, Euthystachys abbreviata* and *Eurylobium serrulatum* are heath-like. The great culmination of the ericoid form is, of course, seen in the Ericaceae themselves, where the huge genus Erica alone has over 450 mountain and south-western species.

4. Among all the dominant shrubs there has been a general tendency towards reduction in size while retaining the woody character. Trees are very rare. With the transition to the shrub form there is the usual increase in branching. The root systems, so far as our knowledge goes, appear to be well developed. They penetrate deeply as well as spread out widely near to the surface.

5. There is a general prevalence of minor xerophytic characters, such as hairy or woolly coverings to leaves, thick cuticle, sunk stomata, etc. Pubescence is particularly a feature of the mountain representatives, e.g. numerous species of *Helichrysum,* but also of some of the Proteaceae (species of *Protea, Leucadendron argenteum,* etc.) and some of the Leguminosae (*Podalyria*), as well as others. Such xerophytic characters, however, though common enough, are not so characteristic as the other features mentioned.

Some of the most interesting of the negative features of the

Fynbosch trees and shrubs are best brought to light by comparison with xerophytic eastern types. Succulence, that common feature of the vegetation of dry regions, is not characteristic of the south-western flora. Such succulent types as do occur are largely to be considered outliers of the Karroo or eastern subtropical flora. Thorn development is also more or less characteristic only of eastern outliers in the south-west, e.g. species of *Gymnosporia*, *Asparagus*, *Solanum*, *Lycium*, but a few of the Macchia shrubs have prickly leaves, e.g. *Cliffortia* spp., and among the Leguminosae species of *Lebeckia*, *Viborgia* and *Aspalathus* are often more or less spiny. *Mundia spinosa* is common. Altogether probably not more than 2 or 3 per cent of the Macchia shrubs are spiny, a very low percentage considering the xerophytic nature of the vegetation. The compound type of leaf, which, as we have seen, is found in over 40 per cent of eastern tree-veld and scrub species, is also rare in the south-west. A section of the genus *Cliffortia*, a few south-western genera of Leguminosae (e.g. *Cyclopia*, *Lebeckia*, *Virgilia*, *Psoralea*, *Crotalaria*), species of *Rubus*, *Peucedanum*, *Pelargonium* and *Euryops* show it as well as outliers of eastern genera, e.g. *Rhus*, but on the whole it is not much more common than thorn development.

Though there is considerable diversity in detail among the mountain and south-western shrubs, there is, on the whole, a wonderful uniformity in the general sclerophyllous type. The distribution of this sclerophyllous vegetation is, to a large extent, determined by the seasonal rainfall. The total annual rainfall is not small – in fact, portions of the south-western region are the wettest in South Africa – but the rain falls almost entirely during the winter months, and there are several months in the hottest summer season more or less without rain. The tallest, most mesophytic Macchia shrubs and the areas of forest occur, however, in places where the summer heat is tempered by the south-eastern mist-clouds. From these, as Marloth has shown, the deposition may be considerable, though it does not fall in such a way that it can have any effect on an ordinary rain-gauge. It drenches the vegetation, however, with moisture. On the Drakensberg the climate differs. The dry season is the winter season of low temperatures. In summer rain falls at fairly frequent intervals. Yet there are also intervals of drought in summer, and these may be rather intense. The high altitude increases evaporation during clear weather, but mist-clouds are again frequent. Though there is a marked difference in the seasonal distribution of rainfall between the eastern mountains and the south-western region, the climates of the two regions probably have more in common than at first sight might appear to be the case.

The uniformity of growth-form type in the whole temperate flora of South Africa is perhaps partly to be explained by its common origin. The dominant species belong to relatively few, though unrelated, families. At the same time, it is remarkable how widely diverse phylogenetically such a type as the heath form really is, including, as it does, a large number of the subordinate associated species.

The south-western flora of South Africa resembles closely that of the Mediterranean region ecologically. I have used the terms Macchia and Fynbosch more or less as synonymous, but the South African Fynbosch is sufficiently distinct to justify the retention of the local name. Cooper [4] finds that the Chaparral of California is also ecologically equivalent to Macchia. His detailed analysis and descriptions of Chaparral show that it has exactly the same main characters as the South African Fynbosch. It is dominated by species belonging to genera unrelated taxonomically, the most important features of which are the root system, extensive in proportion to the size of the plant, the dense, rigid branching, and pre-eminently the leaf, which is small, thick, heavily cutinised and evergreen.

REFERENCES

[1] SIM, T. R. (1907) *The Forests and Forest Flora of the Colony of the Cape of Good Hope* (Aberdeen).
[2] BEWS, J. W., and AITKIN, R. D. (1923) 'Researches on the vegetation of Natal', *Union Botanical Survey Memoir*, v.
[3] ENGLER, A., and PRANTL (1889) *Die natürlichen Pflanzenfamilien*.
[4] COOPER, W. S. (1922) 'The broad-sclerophyll vegetation of California', *Carnegie Institute of Washington*, no. 319.

11 Floristics and Ecology of the Mallee

J. G. WOOD

From 'Floristics and ecology of the mallee', *Trans. Roy. Soc. S. Aust.*, LIII (1929) 359–66.

THE present paper is the result of a study of the mallee scrub, extending over six years, and includes observations made in the Murray Mallee (the area in South Australia extending from 37° to 32° S. latitude), Yorke Peninsula, portions of Eyre Peninsula, outliers of the mallee in the north-east of South Australia, and also in the Millewa and Wimmera districts of Victoria and the mallee regions north of the river Murray in New South Wales (Fig. 11.1).

The term 'mallee' is a native one and refers to a habit of growth. The first published record of the name is in *Australia Felix*, published in 1848 by W. Westgarth, who states that 'the natives of the Wimmera prepare a luscious drink from the *laap*, a sweet exudation from the leaf of the mallee (*Eucalyptus dumosa*)'. The term, however, was probably in common use amongst the settlers at this period.

Mallee scrub is typical of hundreds of square miles of country in southern Australia, the chief tree covering being various species of the genus *Eucalyptus*. The most important species are *Eucalyptus dumosa*, *E. oleosa* and *E. gracilis*, while more locally are found *E. incrassata*, *E. calycogona*, *E. angulosa*, *E. leptophylla*, *E. cneorifolia*, *E. flocktoniae*, *E. behriana*, *E. diversifolia* and *E. pyriformis*. All these species are small trees varying in height from about 2 to 12 m. and all with a characteristic habit of growth. From a main underground root-stock several stems arise which branch sparingly and bear leaves only at the ends of the branches. This results in the formation of a canopy-like collection of leaves. The coppiced habit and canopy top are the two outstanding features which give the facies to the whole of the area.

Associated with the *Eucalyptus* spp. are commonly found the 'native pines', *Callitris robusta*, *C. verrucosa*, *Casuarina lepidophloia*, and various shrubs and undershrubs of a xerophytic type. The ground flora is largely ephemeral and is present only after rain. For the

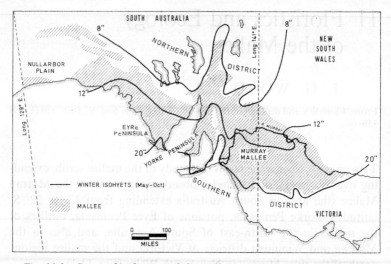

Fig. 11.1   Geographical range of the mallee in part of southern Australia

great part of the year the ground beneath the trees and shrubs is bare (see Plate 11(*a*)).

Practically no previous work relating to the ecology of the mallee has been published, though floristic lists have been made by Tate [1, 2, 3] and Tepper [4] in South Australia, and by Hardy [5] in Victoria. Adamson and Osborn [6] studied a mallee community at Ooldea.

VEGETATION

Floristics

The uniformity in habit of the chief tree species of the mallee scrub has obscured the true nature of this formation. It has been regarded as a formation comparable to the Eucalyptus forests of the Mt Lofty Ranges, so ably dealt with by Adamson and Osborn, or the communities of the arid flora. Analysis of the flora of the mallee, however, shows that it is a great transition region containing elements of both the Southern and the Northern Regions, and with but few endemic species. The different species of Eucalyptus which make up the mallee are co-climatic species in a great transition region, but the chief shrubs and undershrubs belong to other communities. It might be well to revive the old terms of Tate [7] and designate the Northern

Flora the Eremian Flora, and the Southern Flora the Euronotian Flora, with its chief centre of distribution in South Australia in the Mt Lofty Ranges. The dividing line between the two is the 12-in. rainfall isohyet.

The area occupied by the mallee is a stable one geologically, and apparently floristically as well. Few communities can be recognised in it, and these are usually conditioned by slight edaphic differences and seldom occupy any considerable area.

The following Table 11.1 gives the analysis of the total mallee flora in South Australia:

Table 11.1

		per cent
Total number of species	590	–
Number common with Southern Region	194	35
Number common with Northern Region	204	35
Wides (shared with North and South)	27	4·5
Total number of species shared	425	74·5
Number confined to region	165	25·5

From this table it is evident that of the total number of species found, only 25·5 per cent are confined to the region, and that the great majority are plants from other regions. Of these confined plants the chief families and genera are:

Acacia	17 species	} Leguminoseae – 32 spp.
Papilionatae	15 ,,	
Compositae	19 ,,	
Eucalyptus	9 ,,	} Myrtaceae – 18 spp.
Other Myrtaceae	9 ,,	
Chenopodiaceae	12 ,,	
Proteaceae	7 ,,	
Sapindaceae	7 ,,	
Goodeniaceae	7 ,,	

The migrants from the Southern Region are chiefly members of the Savannah Forest [8].

Further insight into the composition of the mallee flora can be gained by treating separately the three great territorial elements – the Murray Mallee, Yorke Peninsula and Eyre Peninsula.

1. *The Murray Mallee.* The analysis of this region is given in Table 11.2:

Table 11.2

		per cent
Total number of species	490	–
Number common with Southern Region	152	32
Number common with Northern Region	183	37
Wides	22	4·5
Total number of species shared	357	73·5
Number confined to Region	133	26·5

In this region the number of species found is greater than in either of the other two, and of these the greater number is shared with the North. Of those confined to the region 28, or 6 per cent, are endemic, that is to say not shared with Yorke Peninsula or Eyre Peninsula. These plants are: *Lamarkia aurea*, *Codonocarpus pyramidalis*, *Scleranthus minisculus*, *Lepidium monoplocoides*, *L. dubium*, *Acacia rhetinocarpa*, *A. montana*, *A. bynoeana*, *A. menzellii*, *Pultenaea prostrata*, *Dillwynia uncinata*, *Beyera opaca*, *Dodonaea cuneata*, *Spyridium subochreatum*, *Cryptandra amara*, *Pimelea williamsoni*, *Callistemon brachyandrus*, *Eucalyptus incrassata*, *Micromyrtus ciliata*, *Halorrhagis ciliata*, *Eremophila divaricata*, *Velleia connata*, *Dampiera marifolius*, *D. rosmarinifolius*, *Olearia hookeri*, *Cassinia aculeata*, *Humea pholidota*, *Calocephalus sonderi*.

2. *Yorke Peninsula*. Table 11.3 shows the analysis of the flora of this region.

Table 11.3

		per cent
Total number of species	302	–
Number common with Southern Region	142	47
Number common with Northern Region	72	24
Wides	24	8
Total number of species shared	238	79
Number confined to region	64	21

Endemic species (not found in the Murray Lands nor on Eyre Peninsula) are as follows: *Prasophyllum fuscoviride*, *Hibbertia billardieri*, *Phebalium glandulosum*, *Veronica plebeja*, *Eucalyptus behriana*, *E. flocktoniae*.

Shared with Eyre Peninsula, but not found in the Murray Lands, are: *Grevillea aspera*, *Didymotheca thesiodes*, *Acacia acinacea*, *Daviesia genistifolia*, *Phyllanthus calcynus*, *Lasiopetalum discolor*,

Thryptomene miqueliana, Leucopogon woodsii, Scaevola humilis, Olearia exiguifolia and *Angianthus phyllocalymmeus.*

Compared with the Murray region, the oustanding feature of the region is the large number (47 per cent) of plants shared with the Southern Region and the small number shared with the North. This is to be expected when its geographical position is considered.

3. *Eyre Peninsula.* The analysis of the flora of this region follows in Table 11.4.

Table 11.4

		per cent
Total number of species	406	–
Number common with Southern Region	150	36
Number common with Northern Region	100	25
Wides	23	5·5
Total number of species shared	273	76·5
Number confined to the region	133	23·5

Endemic species are: *Eragrostis trichophylla, Poa drummondii, Triodia lanata, Loxocarya fasciculata,* Caladenia toxochila, Hakea cycloptera, Chenopodium triangulare, Hemichroa diandra, Calandrinia brevipedata,* Hutchinsia drummondii, Acacia gonophylla,* A. merrallii,* A. cyclopis,* Bossiaea walkeri,* Templetonia battii, Gastrolobium elachistum, Dicrastylis verticillata, Solanum hystrix,* Anthocercis anisantha,* Eremophila parviflora,* Helipterum humboldtianum,* H. haigii,* H. tenellum,* Humea assiniifolius.*

This region shows migrants from another flora, for of these endemics those marked with an asterisk, 11 in number, are West Australian species.

Compared with the Murray Mallee, this region is characterised by the relatively small number of species shared with the North. This is the same as the number of Yorke Peninsula species shared with the North. This gives a basis for the fact noted by Prescott, namely, that the division between mallee and mulga country is sharper than the division between mallee and saltbush. The saltbush communities are typical of the north-east of South Australia and the mulga communities of the west of South Australia.

The southern limit of saltbush is shown on Prescott's vegetation map [9]. This limit was compiled from survey data, but it follows closely the 12-in. isohyet. This line has a certain historical interest. In 1865, consequent on the great drought culminating in that year, the Surveyor-General of South Australia (G. W. Goyder) was ordered to 'determine and lay down on the map, as nearly as practicable, the line of demarkation between that portion of the country where rainfall has extended and that where the drought prevails'.

Goyder reported that the 'line of demarkation extends from Swan Reach and then in a north-westerly direction to Burra Hill, then to Ulooloo and Mount Sly, and in a north-westerly direction by Tarcowie and Mount Remarkable, thence south by Ferguson's Range to the Broughton, and south-west to the east shore of Spencer Gulf to Franklin Harbour, and then north-west to the west of the Gawler Ranges'. This is approximately the 12-in. isohyet, and is the boundary between the Northern and Southern Regions.

Analysis of the growth forms

In Table 11.5 an analysis of the life forms as defined by Raunkier is given. These are given as percentages. In the first horizontal column are given the usual symbols for the life form class. These are:

Phanaerophytes: plants with dominant buds exposed freely to the air.

MM *Mesophanaerophytes*, trees; 8–30 m.

M *Microphanaerophytes*, small trees and shrubs; 2–8 m.

N *Nanophanaerophytes*, shrubs; 2 m. and less.

Chamaephytes Ch: buds perennating on surface of ground or just above it (25 cm.).

Hemicryptophytes H: dormant buds in upper soil.

Geophytes G: dormant parts well buried.

Therophytes Th: ephemeral plants – annual plants with short life-cycle.

Epiphytes E: These, in South Australia, are parasitic plants.

Table 11.5

	MM	M	N	Ch	H	G	Th	E
Normal	6	17	20	9	27	3	13	3
Mallee	–	13	30	19	11	3·5	21	–
Murray Mallee	–	13	31	19	12	2·5	21	1·5
Yorke Peninsula	–	13	32	15	9	6·5	22	1·5
Mallee (indigenous)	–	21	40	16	8	1	14	–
Savannah forest [8]	1	8	14	12	26	23	14	1·5
Stringybark forest [8]	1	9	34	13	23	13	4	1
Ooldea [6]	–	19	23	14	4	0·5	35	4

The spectrum for the total mallee region is given, also that for the mallee of the Murray Mallee and for Yorke Peninsula. For comparison the normal or 'world spectrum' is given, also that for the forests of the Mt Lofty Ranges and that for the arid communities at Ooldea.

The outstanding feature of the mallee spectrum, as compared with the normal, is the high percentage of small woody plants (Nanophanaerophytes) and of ephemeral plants or Therophytes. It is a sclerophyll formation with a marked ephemeral element. Compared with the spectra for the savannah forest and Ooldea, it is seen that the mallee is intermediate between the two. The decrease in the number of geophytes characteristic of the savannah is marked, whilst the increase in the therophytic flora is striking but is not as high as that of Ooldea.

Comparing the spectra for the Murray Mallee and for Yorke Peninsula with its large percentage of southern forms, the most marked difference is in the percentages of geophytes, which are relatively high under the higher rainfall of Yorke Peninsula.

An analysis is also given of the plants indigenous to the mallee. This is interesting, as it shows clearly the importance of the shrubs and woody undershrubs which together make up 76 per cent of the plants characterising the region. The percentage of therophytes for these indigenous plants is the same as that of the savannah forest, and the high percentage for the total mallee shows that this is largely due to the northern forms – chiefly species of Compositae and Cruciferae.

REFERENCES

[1] Tate, R. (1879) *Trans. Phil. Soc. S. Austr.*, II 118.
[2] —— (1880) *Trans. Roy. Soc. S. Austr.*, IV 137.
[3] —— (1889) ibid., XIII 112.
[4] Tepper, J. G. O. (1879) *Trans. Roy. Soc. S. Austr.*, III 25.
[5] Hardy, A. D. (1914) *Vict. Naturalist*, XXX 148.
[6] Adamson, R. S., and Osborn, T. G. B. (1922) *Trans. Roy. Soc. S. Austr.*, XLVI 539.
[7] Tate, R. (1890) *Flora of Extra-tropical South Australia* (Adelaide).
[8] Adamson, R. S., and Osborn, T. G. B. (1924) *Trans. Roy. Soc. S. Austr.*, XLVIII 87.
[9] Prescott, J. A. (1929) *Trans. Roy. Soc. S. Austr.*, LIII 7.

12 The Thickets and Woods of the Mediterranean Region

H. HARANT and D. JARRY

Translated by S. R. Eyre from *Guide du naturaliste dans le Midi de la France.*
II. La Garrigue, le maquis, les cultures (Delachaux et Niestlé, Neuchâtel, 1967)
pp. 171–83.

> My kids with their mothers wander safely in
> the forests, there to browse upon the thyme and the
> arbutus, and they fear neither the snakes nor the
> cruel wolves.[1]
>
> HORACE

WITH the habitual exaggeration of the people of the Midi, where a
hill of 600 m. is a 'mountain' or a 'peak', a 'forest' is often no more
than a thicket or the remnants of a wood! For, in spite of its importance in historical times, real forest is rare in this Mediterranean region.
The area actually occupied by evergreen oak woodland is trifling as
compared to that of its theoretical former extent. Roussillon,
Languedoc and western Provence are in the evergreen oak zone,
but the vegetation here is very much degraded. In eastern Provence
this zone fades out, the holm-oak being mixed with deciduous oak,
hornbeam, Aleppo pine and maritime pine. On the one hand
Quercus ilex is everywhere dominant; on the other it is everywhere
subordinate to pine.

At the time of the establishment of the Azilian sea-level, when the
atmosphere was mild and humid, pioneer forest species invaded the
land – the pubescent oak, the Montpellier maple and the Spanish
broom. Then, as the temperature rose again, the appearance of the
landscape changed. New species established themselves – the evergreen oak, filaria alaterne (buckthorn), mastic and terebinth,
accompanied by climbing plants. The undergrowth harboured
numerous thorny scramblers such as the asparagus and butcher's
broom.

At the present time this struggle for supremacy still goes on. The
pubescent oak is relinquishing its hold everywhere; xerophilous and

---

[1] This quotation from Horace does not appear in the latest (1967) edition
(Ed.).

heliophilous Mediterranean plants are establishing themselves with greater ease – particularly the Kermes oak, which is able to ascend to 500 m. above sea-level. The evergreen oak, tolerant of a range of soil types but most successful where there is underlying solid rock, can invade even higher. Deciduous oak is retreating; it holds its own with most success in basins where the soil is less skeletal; unfortunately it has had to withstand the onset of man ever since the Iron Age.

As we have said, only vestiges of the primeval forest remain. Prehistoric man here made clearings and openings for his settlements. Roman villas (name-endings in -*an* and -*ac*) cleared land here for cultivation and for pasture. The intensive and probably profligate exploitation of wood and its clearance by fire has reacted in favour of species which, on the one hand, are adapted to high temperatures and high light intensities and, on the other, to frequent firing. Medieval glass factories, of which some unusual remains are to be found at La Boissière, helped to turn much of the garrigue here into an 'Arabia Petraea' (to use the words of Michelet).

The sclerophyllous forest of Languedoc is characterised by trees and shrubs which are never very tall and which have branched and writhen trunks. Their leaves, shiny and smooth, have thickened main veins and generally well-developed vascular systems.

The holm-oak is the important afforestation species – the only one in this area – but it has been in decline since tanner's bark was superseded by chemicals, and charcoal by paraffin. It is found in copses of low yield, sometimes hardly producing more than 6 cu. m. of bark per hectare (a little more with good management), but actual felling is hardly more profitable. Having become useless, the tree is neither valued nor tended. If one wishes to see a landscape such as this in an almost natural state, such as existed here a century or two ago, one must seek it in northern Catalonia or on the Middle Atlas.

The undergrowth of a wood of holm-oak, according to P. Duvigneaud, has the following floristic composition. Well-known evergreen shrubs are found here – the filaria, the wayfaring-tree, the buckthorn, the lentisk and the terebinth, as well as climbers such as the honeysuckle, the sarsaparilla and the clematis which persist in open spaces. As for the plants of the undergrowth, they too are very often sclerophyllous and evergreen; this is true of *Ruscus aculeatus, Asparagus acutifolius, Hedera helix, Rosa sempervirens, Asplenium adiantum-*

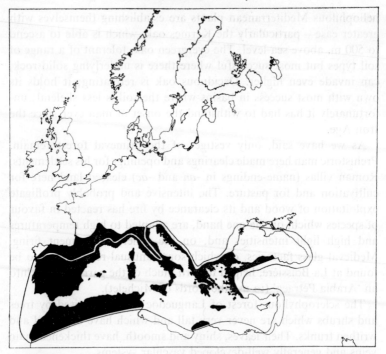

Fig. 12.1   Distribution of Quercus ilex *(after Rikli). The continuous line indicates the limit of the Mediterranean region (according to Braun-Blanquet).*

nigrum, *Teucrium chamaedrys, Viola scotophylla, Vinca major, V. difformis* and *Euphorbia characias.*

This plant association (*Quercetum ilicis*) has two main variants: one is calcicolous with the lentisk predominant, the other is found on siliceous materials, particularly well developed in western Provence, where the deciduous oak thrives.

Since the garrigue cannot be abandoned in its present state, the question has arisen as to how one can ensure the reforestation of soils which are so degraded. It seems that there are several possible solutions all of which are being tried.

On the one hand, one can consider the encouragement of native trees as, for example, by sowing acorns; this was done at Bédoin at the foot of the Ventoux round about 1860. In place of the Aleppo pine and the umbrella pine, which burn too easily, the indigenous cypress has been favoured, whose wood has a certain value in paper-making. This experience in reforestation, using the local species, has been turned to good account in several places; it is much in evidence at Fontfroide (Hérault) and at Mérindol (Vaucluse).

On the other hand, exotics have been introduced by the Service des Eaux et Forêts, particularly the Moroccan cedar and the Arizona cypress (*Cupressus arizonica*). The first, which demands little of the soil, comes down to fairly low altitudes; it is more important than its related species from the Lebanon, Cyprus and the Himalayas which appear to give a smaller economic return. The second, only introduced in 1930, appears to have better characteristics than Lambert's cypress (*Cupressus macrocarpa*) which lives for only a short time.

Generally speaking, the forests of the Côte d'Azur can be divided into two main associations, the one on calcareous and the other on siliceous rocks. The first is represented by the Aleppo pine forest at low altitudes between Cannes and Menton. The under-storey is dense and invested with lianes which climb up to the tops of the pines. The species-content is always poor.

In order of importance we have: *Myrtus communis, Rosmarinus officinalis, Pistacia lentiscus, Juniperus oxycedrus, Phillyrea angusti-folia, Calycotome spinosa, Rhamnus alaternus, Daphne gnidium, Arbutus unedo, Erica arborea* and bush *Quercus ilex.*

On the very skeletal soils on hard limestone, less common plants are found in association: *Osyris alba, Cneorum tricoccum, Ceratonia siliqua* and *Olea oleaster.* The carob tree and the wild olive make one wonder precisely what it is which debilitates a neighbouring group of the *Oleo–Ceratonion*, which is the forest climax of the southern Mediterranean and Spain.

The siliceous massifs of the Maures and the Estérel were formerly more wooded than they are today. Fire has transformed them into a treed maquis with evergreen oak, deciduous oak, cork oak, Aleppo pine, maritime pine and umbrella pine. Some fine remnants of this type of forest, with a dense under-storey of arbutus, survive on the sandstones of Menton – at the Annonciade and near the Sanatorium of Gorbio. The profits from plantations of the holm-oak are great, but the components of the undergrowth are always distinctly acidophile: *Erica arborea, E. scoparia, Cytisus triflorus* and *Lavandula stoechas.* Certain montane species are found here below their usual level: *Rhus cotinus, Pteris aquilina* [*Pteridium aquilinum* (Ed.)], *Cytisus sessiliflorus, Genista hispanica, Aster sedifolius acris* and *Osmunda regalis.*

In several places one particular component becomes predominant:

on the north of the Maures it is the umbrella pine, in the Estérel
the maritime pine and, at Saint-Raphael, pines, Spanish juniper and
a ground cover of *Cladonia*.

THE CHIEF TREES OF THE OAK AND PINE
PLANTATIONS

The evergreen oak (*Quercus ilex*; its systematic name stems from
the two Latin nouns for the oak and the holly) belongs to the
Fagaceae along with the beech and the chestnut. In the garrigue this
tree is seldom more than 15 m. high, but old specimens, which are
unusual elsewhere, survive in places which are not easily accessible
or in reserves; they can achieve 20–25 m. (as in the old part of the
Jardin des Plantes at Montpellier where there are fine examples
which date back to the end of the sixteenth century).

Its brown bark is delicately fissured. Its greenish-grey foliage gives
the countryside that wasted appearance which characterises areas of
garrigue. Whether as a bush or a tree, it is indeed king in this land and
its distribution falls little short of covering the whole of the Mediterranean region (Fig. 12.1).

The young branches are downy; the evergreen leaves, leathery and
corrugated, are extremely variable in shape and size, often on the
same tree. They may be ovato-lanceolate and more or less narrow;
they may be elliptical or almost orbicular; they may be dentate and
spiny when young, entire and almost free of spines on old individuals
and old branches; they have green upper surfaces and are greyish
beneath. The polymorphic acorns are half enclosed in a cup whose
base is rounded and which has adpressed scales.

The evergreen oak, which is adapted to the poorest land and can
survive prolonged periods of drought, is also remarkable in its ability
to regenerate from basal sprouts and suckers. In some areas, notably
away from the coast, it forms mixed populations with the pubescent
oak and, here and there, it hybridises with the Kermes oak to produce
Q. auzandi. Several Languedoc surnames such as Euzet, Euzière,
Deleuze and Dieuzède are derived from the name of this tree.

The cork oak or 'surier' (*Q. suber*) resembles the evergreen oak in
its evergreen though less luxuriant foliage; it is always distinguished
by its more deeply fissured bark, however, and by its conical acorn
cup with raised scales arranged in continuous delicate bands. It is
markedly calcifuge and is important only in Provence, in the Maures

and the Estérel, and in Roussillon; but it has been planted in many places. Its height is 10–15 m.

Fig. 12.2   Distribution of Quercus pubescens *(after Zôlyomi)*

The 'Rouvre', pubescent oak or white oak (*Q. pubescens lanuginosa*) has a twisted trunk 15–25 m. in height and blackish bark with short, deep fissures. The young branches are hairy and whitish; the leaves, downy when young, and remaining so on the underside for a long time, are also more or less deeply lobed. The fruits are clustered and almost sessile; they are borne in cups with suppressed, hairy scales. This oak is a tree which inclines to the montane habitat but it is found downslope almost into the plain. It thrives on the relatively young siliceous alluvium of Pliocene age (Fig. 12.2). In some basins in the garrigue where ponding of cold air takes place in winter, the evergreen oak and pubescent oak zones are inverted, the latter being at the centre with the evergreen oak above it. This can be seen very clearly at Saint-Martin-de-Londres.

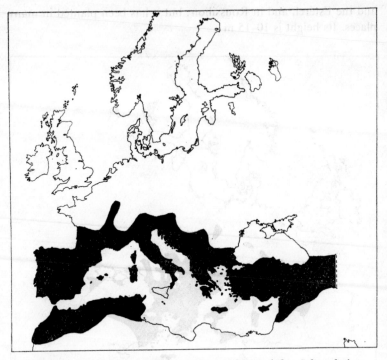

Fig. 12.3   Distribution of Acer monspessulanum *(after Schmucker)*

The Montpellier maple (*Acer monspessulanum*), which is often mixed with evergreen oak, is according to Braun-Blanquet an element of the shrub stratum in the evergreen oak forest (Fig. 12.3). It never grows to a height of more than 5–6 m. Its bark is a yellowish grey, smooth at first but developing longitudinal cracks. The leaves, with their distinctive shape, are smooth and glaucous beneath, with only three equal lobes which are entire or only slightly sinuate, the sinuations forming almost right-angles with each other. The greenish-yellow flowers, arranged in drooping clusters, appear before the leaves. The fruits are samaras with inflated fruit coats which extend into wings; these converge in the form of pincers and are narrower at the base.

The Aleppo or Jerusalem pine (*Pinus halepensis*) can attain a height of 15–16 m.; it can easily be distinguished by its sinuous trunk and rounded crown. The ash-grey bark is smooth in young individuals and fissured in mature trees. The branches are short, tapering and

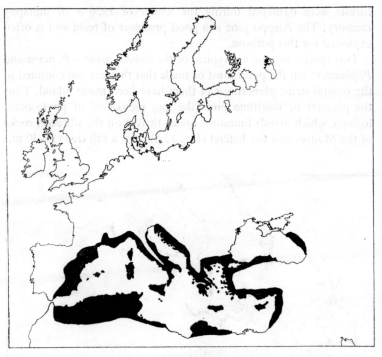

Fig. 12.4   Distribution of Pinus halepensis (*after Schmucker*)

spreading; the leaves, which are 9–16 cm. long, are soft and paired (rarely in threes or fives); the fruits or cones, 6–12 cm. long, pedunculate and always reflexed, are elongated, narrow towards the tip, and truly conical. The scales of the latter, tawny brown or reddish, have their tip or boss almost flat, faintly keeled and armed in the middle with a blunt, scarcely projecting point. The blackish seeds, 7 mm. long, have a wing four times as long as themselves.

The Aleppo pine, along with the evergreen oak, is the most widely distributed tree on limestone areas in the Midi (Fig. 12.4); here it is very numerous, particularly in Provence but also in Languedoc as far as the outskirts of Montpellier. According to Braun-Blanquet 'it is only a pioneer species; it can gain a foothold in eroded soils which repel all other species and, once established, it improves the microclimate by reducing its dryness'.

Its rate of growth is rapid, but it suffers from the great limitation of being very susceptible to bad frosts, and a large number of pine forests were damaged during the winter of 1956 – of unhappy memory! The Aleppo pine is a good producer of resin and is often exploited for this purpose.

Two species of pine are typical of the coastal areas – *P. pinea* and *P. pinaster*; but the point must be made that they are not confined to the coastal strip: plantations of them have been made inland. Thus the pinaster or maritime pine (the 'pin mésogéen' of phytosociologists), which avoids limestone areas, thrives on the siliceous rocks of the Maures and the Estérel (Fig. 12.5). It is a tall tree (15–30 m.)

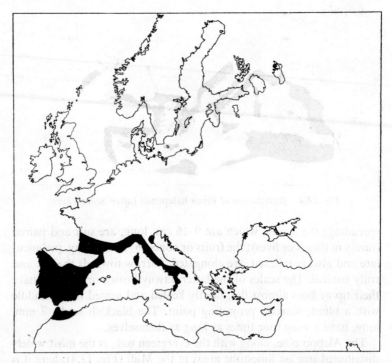

Fig. 12.5   Distribution of Pinus pinaster (*after Schmucker*)

with a straight trunk and pyramidal crown. The needles are rigid, thick and long, being flat above and convex beneath. The cones, larger than those of the Aleppo pine, are 12–18 cm. long: they are expanded or reflexed-acuminate at the tip; their scales are a shiny russet yellow, the bosses being distinctly keeled with a smaller central, pyramidal pimple; the seeds, 9–10 mm. long, have wings four or five times as long as themselves.

The umbrella pine (*P. pinea*), that much-valued feature of the landscape planted in so many parks and gardens, is distinguished from the preceding species by the spreading crown which gives it so picturesque an appearance. Though not native, it is common everywhere in the south of France and is an abundant species in the littoral zone (as in the peninsula of Giens and on the old dunes of Camargue, Aigues-Mortes and Grau-du-Roi) as well as inland (Fig. 12.6).

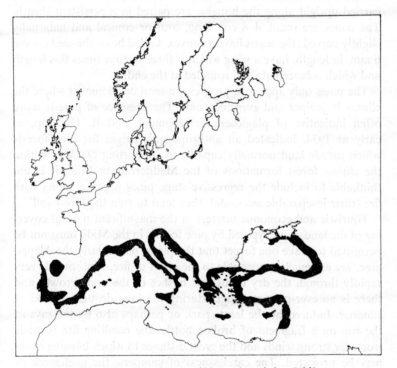

Fig. 12.6   Distribution of Pinus pinea (*after Rikli*)

The leaves, arising from a persistent sheath, are pointed and prickly and more than 16 cm. long. The cones are 15 cm. long and 10 cm. in width with large scales which are furrowed with cavities on the inner face where the seeds are lodged; these have a hard testa which is oblong and about 20 mm. long and which encloses the edible kernel or 'pignon' used in pastry. The boss of the scales is pyramidal with a short point and poorly developed transverse keel.

The other kind of pine which is found here is the Salzmann or Cévennes pine (*P. nigra salzmanni*). It is considered to be a local variety of the well-known Corsican pine and is endemic on the Cévennes limestones (in Hérault, Gard and Ardèche) and in the Pyrenees. It forms a fine and famous forest at Saint-Guilhem. This pine is low-growing (6–10 m.) and much twisted, with horizontal branches and a crown which is very often completely flattened. The scales of the bark are a silvery grey. The leaves, 12–15 cm. long and carried upright along the boughs, are paired in a persistent sheath. The cones are small, 4–8 cm. long, oval or conical and habitually slightly curved; the scales have a convex, keeled boss; the seeds, some 6 mm. in length, have a wing which is three or four times this length and which is lanceolate and rounded at the end.

The pines only appear in progressive seral development where the climax is juniper and evergreen oak. The presence of pine is more often indicative of plagiosere development, and R. Ducamp, as early as 1934, indicated an anthropogenic origin for pine forests which invade land normally capable of supporting *Q. ilex*: 'within the climax forest formations of the Mediterranean Midi, it is unthinkable to include the regressive stage pines because, along with fire (their inseparable associate), they tend to ruin the forest soil'.

Touristic and economic interest in the magnificent natural covering of the landscape, typified by pine forests in the Midi, must not be permitted to make one forget that the pines, particularly the Aleppo pine, are extremely susceptible to the effect of fire. Fire spreads very rapidly through the dry herbs and bushes of the undergrowth and there is an ever-present danger during the long drought period of summer. Induced by the least spark or perhaps also by the rays of the sun on a fragment of broken bottle, the resulting fire is made worse by strong winds and the great distance to which burning cones may be projected. The carelessness of campers, the negligence of smokers, sometimes even premeditated mischief, are the cause of these calamities which only a strict discipline would be able to prevent.

Fire has become a common sight in the south of France and in Corsica; many local inhabitants and holidaymakers are now familiar with its ominous roar, its heavy smoke plume and its smell of burning resin. When the mistral is blowing, this scourge fans out and one can watch it advancing in the Maures at 100 km/h. In the Mediterranean region, the blanket of ash has no fertilising power; it sterilises seeds, takes up water and destroys humus. Wildlife is decimated: for each hectare that is burned it has been estimated that 300 birds, 400 mammals, 100 tortoises, snakes and toads, and 5 million insects disappear.

The struggle is particularly difficult in country so hard to penetrate because of its cliffs and ravines, and also because of the scarcity of surface water. Catalina or Martin flying boats are used to spray tons of sea water when houses are threatened. Despite the extent of the resources deployed, 25,000 ha. have been burned in the Maures in the first week of August 1965.

13 Grassland Climax

CARL O. SAUER

From *Agricultural Origins and Dispersals: The Domestication of Animals and Foodstuffs* (M.I.T. Press, Mass., 1969) pp. 15–17.

THE young science of ecology has undertaken to study the associations of organisms, initially as belonging together by their physiologic requirements or their joint adaptation to a particular physical environment. Systems of classification arose that identified plant and animal complexes with climate. Thus there arose the concept of the 'ecologic climax', currently defined as 'the final or stable type of plant community reached in a particular climate'. A postulate tends to displace reality. Climatic regions are cartographic abstractions, useful as elementary teaching devices to give some first notions of weather contrasts over the earth. 'Final or stable' communities are quite exceptional in nature: weather, soils and surfaces are continually changing; new organisms are immigrating or forming, old ones may be giving way. Change is the order of nature: climax assumes the end of change.

Since there are many grasslands about the world, such associations have been inferred to represent a grassland climax, that is, to be stable and the result of particular climates. The theme song became that every single grassland must have a climate for its own. When I had ecologic instruction with the admirable Cowles at Chicago, I was at first persuaded. More and more I had to admit that I was unable to find the coincidence of grassland and limiting values of moisture. Why should heavy-rooted woody plants be excluded from certain areas and these surrendered to the exclusive occupation by herbs and grasses? Thus I began to surmise that the plant ecologists had construed away the role of animals and especially of man.

I grew up in the timbered upland peninsula formed by the junction of the Missouri and Mississippi rivers. The prairie began a few miles to the north and extended far into Iowa. The broad rolling uplands were prairie, whatever their age and origin, the stream-cut slopes below them were timbered; river and creek valleys and flanking ridges were tree-covered, be they formed in bed rock or on deep loess mantle. From grandparents I heard of the early days when people dared not build their houses beyond the shelter of the wooded slopes, until the plough stopped the autumnal prairie fires. In later field work in Illinois, in the Ozarks, in Kentucky, I met parallel conditions of vegetation limits coincident with break in relief. I gave up the search for climatic explanation of the humid prairies.

The Far West, of subhumid and semi-arid climates, showed again the same relation of grassland to topography. The grassy valley basins of California, bordered by oak parks of senescent, non-reproducing trees [1], the high plains of Arizona, New Mexico, Chihuahua and Durango all are or had been grasslands lapped against rough, woody country lying above or below the plains. In the Latin-American vernacular, *monte* (mountain) has come to mean brush or woodland. Last, I became acquainted with tropical savannas in Central and South America, plains often studded with tall palms, with gallery forests along the valleys and with woody growth wherever there is broken terrain. Parts are semi-arid, parts get as much rain as any part of the Mississippi Valley. Cattle grazers still keep up the aboriginal burning practice.

Such, in brief, has been the sum of my experience: grasslands occupy plains; woody growth dominates rough terrain. It does not follow, of course, that plains must be grassy, though grasslands and pine woods do characterise a lot of them. Fires sweep most freely over smooth surfaces, spreading before the wind until they are stopped at the brink of valleys cut below, or die out in hill and mountain lands that rise above them.

Grasslands shrink where protected from fire. The wood's margin may advance by shading out the grasses. Animals and wind carry seeds, some of which grow where accidental openings occur in the herbaceous cover, and these accidents do happen. In the Kentucky Barrens or Pennyroyal [2] a gently undulating limestone upland of high fertility is surrounded by stream-cut terrain, the latter, at settlement, a forest of mixed hardwoods, the former a grassland, whence its name, Barrens. The Pennyroyal a century ago was noted by the gifted geologist David Dale Owen as having self-seeded woodlots in former grassland. These were enclosures fenced off to serve as stock pens. Owen attributed the grassland to fires, as by Indian burning, the woodlots to protection by white farmers. Aided by grazing stock, there has been under way lately an explosive advance of leguminous scrub, mesquite, catclaw, huisache, over former grasslands in northern Mexico and the South-west. On the Staked

188 *Carl O. Sauer*

Plains of Texas there has been wide invasion by dwarf white oak brush, on the southern Colorado Plateau of sagebrush.

In the natural course a maximum depth of plant growth tends to develop on any site. Between ceiling and floor a maximal diversity of organisms is accommodated to the full utilisation of moisture, light, minerals and organic food. Grasslands are living zones greatly reduced in depth, above and below the ground; they are simplified morphologically, and usually reduced as to diversity. They are an impoverished assemblage, not a fully developed organic household or community.

REFERENCES

[1] COOPER, W. S. (1922) 'The broadleaf-sclerophyll vegetation of California', *Carnegie Institute of Washington*, Publ. no. 319.
[2] SAUER, C. O. (1927) 'Geography of the Pennyroyal', *Kentucky Geological Survey*, series 6, xxv (Frankfort).

14 The Origin of Mixed Prairie

J. E. WEAVER and F. W. ALBERTSON

From *Grasslands of the Great Plains* (Johnson, Lincoln, Neb., 1956) pp. 8–11.

THE work of students of the earth's structure and of plant and animal fossils of past geological periods, together with the study of past climates and climax vegetations, has given us the history of the earth's surface, its animal life, and its vegetation. The origin of Mixed Prairie probably dates back 25 million years to Tertiary times. In the Eocene period a warm temperate forest occupied the Great Plains. At this time the climate was warm and moist. But as the Rocky Mountains rose, they intercepted the moisture-laden winds from the Pacific Ocean. Since water was precipitated mostly to the west of the mountains, only dry winds, which produced very little rainfall, reached the eastern side. Low summer precipitation was accompanied by dry winters.

Clements [1] in his study of Mixed Prairie states: 'It is probable that the evolution of grassland proceeded more rapidly in the period of mountain-making in the upper Oligocene to produce the forerunner of the modern prairie in the Miocene, where the typical genus *Stipa* [needlegrass] is recorded, along with horses of the grazing type, *Merychippus* and *Protohippus*.' During the Miocene 'to the east of the Rocky Mountains, increasing aridity reduced the forest vegetation and its opportunities for preservation to such an extent that the fossil record there is limited. A wide extent of grasslands is indicated by the abundance and diversity of grazing mammals.' Indeed, grazing animals have roamed the Great Plains throughout geological ages. Geologists have found the remains of such ancient animals as the four-toed horse in deposits which they believe to be at least 45 million years old. In fact, the evolution of horses from very small animals approximately to their present size has been traced from their preserved skeletal parts in fossil beds scattered over the plains. Many other mammals long since extinct, as camels, elephants, rhinoceros and primitive types of buffalo, roamed the plains and fed upon the grass and other vegetation. The evidence indicates a great multiplication of species of grasses between early Tertiary periods and the Pleistocene.

'Until recently there was almost no direct evidence of the sedges and grasses which now cover the prairies and form such an essential food for the grazing mammals. Since the teeth of the plains-dwelling mammals showed marked specialisation for such food during the Miocene, it had been inferred that the prairies became widespread at that time. Prairie grass is harsh food and tends to wear down the grinding teeth. To make up for this, modern grazing animals, like horses and cattle, have high-crowned teeth that grow out to replace the loss. All the early Cenozoic mammals possessed low-crowned teeth; but during the Miocene the horses, camels, rhinoceroses, and other groups dwelling on the plains began to show rapid changes to the high-crowned grazing teeth. Within the last decade wonderfully preserved grass seeds have been found to occur abundantly in the sands of middle and late Miocene and Pliocene deposits of the High Plains, and now about 30 species are known. Since none have yet been found in lower beds of the same region, the fossil record now seems to confirm the inference drawn years ago from the adaptation seen in mammalian teeth, namely, that the grasses first spread over the plains in Miocene times' [2]. This discovery in western Kansas and eastern Colorado by Elias [3, 4, 5] included several forms of needlegrass (*Stipa*), a panic grass (*Panicum*) and a bristlegrass (*Setaria*), and several forbs of the borage family.

In Pleistocene time the ice at the period of greatest expansion in the United States covered only a relatively small part of the Mixed-Prairie area in northern Montana and North Dakota but it extended eastward from the Missouri river in the Dakotas. The Great Plains from central Montana and south-western North Dakota southward to Texas were free of ice. The period of glaciation had nevertheless a profound effect upon the cover of vegetation. It seems certain that the Boreal Forest was pushed far southward and later retreated northward during warm, dry periods. Evidence of this is found in the persistence of white spruce to this day at higher altitudes in the Black Hills and the survival of paper birch and aspen, all cold-climate species, in the deep canyons and elevated plateaus. Indeed, many types of vegetation retreated and advanced across the Great Plains during the pulsations of cool, moist and warm, dry periods in Pleistocene time. These mass movements of grassland dominants were apparently associated with a period of intensive evolution.

The eastern edge of the Black Hills has numerous relics of eastern Deciduous Forest trees, such as bur oak, ash, elm and others, which here found refuge and remained behind the eastwardly retreating woodland which once covered the plains. A relic, in an ecological sense, is a community or a fragment of one that has survived some important change, often to become a part of the existing vegetation. Other evidence is the presence of hard maple and other remnants of Ohio forests in the deep Caddo Canyon in west-central Oklahoma. It is also believed that tall grasses occupied the Great Plains for a period following the retreat of the deciduous forest.

According to Clements, an outstanding authority on vegetation, needlegrasses (*Stipa*), Junegrass (*Koeleria*), wheat grasses (*Agropyron*) and wild-rye grasses (*Elymus*) are of northern derivation, as are also various bluegrasses (*Poa*) and Idaho fescue (*Festuca idahoensis*), as well as thread-leaf sedge and needle-leaf sedge (species of *Carex*). All are cool-season plants.

Concerning the short grasses, 'their homeland was the mountain plateaus of Mexico and Central America. Such far-ranging species as [blue grama] *Bouteloua gracilis* and [side-oats grama] *B. racemosa* [*curtipendula*] and [buffalo grass] *Buchloe dactyloides* have pushed north to the Canadian border or beyond, but the majority are confined to the region west and south of Texas. [Three-awn] *Aristida*, although of somewhat different form, is a regular associate of *Bouteloua* and of similar southern origin and xeric nature' [1]. These are distinctly warm-season grasses. Curly mesquite (*Hilaria belangeri*), a dominant short grass of the Texas plains, and galleta (*H. jamesii*) are other grasses of south-western origin.

The warm-season tall grasses, big bluestem (*Andropogon gerardi*), switchgrass (*Panicum virgatum*) and Indian grass (*Sorghastrum nutans*) together with the mid grass, little bluestem (*Andropogon scoparius*), are of semi-tropical origin and are believed to have come from the east and south-east.

With the slow southward movement of the glaciers and increasing cold, wheatgrasses, needlegrasses and others of northern origin were slowly spread into the southern Great Plains and the South-west. Upon the retreat of the glaciers to the polar cap during a long, warm, dry period these grasses (though some remained) migrated far northward. Moreover, the short grasses and their associated warm-season species moved northward from their place of origin on the Mexican Plateau. During one of the glacial epochs (Nebraskan or Kansan) it is believed that certain tall grasses, such as big bluestem, migrated westward ahead of the westerly-moving deciduous forest. Thus, great and repeated migrations and a vast mingling of species over the

Great Plains have resulted from the various pulsations of climate during glacial times.

The dry climatic cycle of 1933 to 1940 has recapitulated in miniature the much larger cycles of long duration of post-glacial times. During these seven years of extreme drought, Mixed Prairie replaced True Prairie over an area of 100 miles in width from South Dakota to central Kansas. This included the destruction of species populations and local mass migrations. Composition of the vegetation was greatly changed and the form of the plants modified by dwarfing. The drought cycle had a marked effect on both vegetation and soil throughout the plains. As usual, the effects were more pronounced along the ecotones or transitional areas from one plant association to another [6].

Deep canyons, steep protected slopes, cool north exposures, sunny dry ridges, and soil with increased moisture, from whatever cause, have been the refuges of numerous species which were able to exist in the changing climates. Such havens were especially numerous in rough country and along the mountain borders.

REFERENCES

[1] CLEMENTS, F. E. (1936) 'Origin of the desert climax and climate', reprint from *Essays in Geobotany* (University of California Press).

[2] SCHUCHERT, C., and DUNBAR, C. O. (1941) *Textbook of Geology*, pt 2, 'Historical Geology' (John Wiley & Sons, New York) pp. 443–4. Reprinted by permission.

[3] ELIAS, M. K. (1932) 'Grasses and other plants from the Tertiary Rocks of Kansas and Colorado', *University of Kansas Bulletin*, XXXIII 333–67.

[4] —— (1935) 'Tertiary grasses and other prairie vegetation from the High Plains of North America', *American Journal of Science*, 5th series, XXIX 24–33.

[5] —— (1942) 'Tertiary prairie grasses and other herbs from the High Plains', *Geological Society of America. Special Papers*, no. 41.

[6] WEAVER, J. E. (1943) 'Replacement of True Prairie by Mixed Prairie in eastern Nebraska and Kansas', *Ecology*, XXIV 421–34.

15 The Steppe and Forest Steppe of European Russia

BORIS A. KELLER

From 'The distribution of vegetation on the plains of European Russia', *J. Ecol.*, xv (1927) 209–18, 227–9.

OAK FOREST

ACCORDING to Morózov the oak is in Russia a species characteristic of the neighbourhood of the steppes. The oak region is a field of battle between forest and steppe, where under natural conditions the forest will obtain the upper hand. In accordance with this view the whole oak region of European Russia may be divided into three zones:

1. The belt of old forest steppe or of continuous oak forest where the struggle has been brought to a conclusion with the victory of the forest, and the loess (earlier steppe soil) is occupied by it. This forest zone is now split into fragments owing to the extensive destruction of the forest by man.

2. The belt of the existing forest steppe where the oak is developed in island-like clumps, near the banks of rivers.

3. The steppe belt, where forest is confined to ravines and does not extend to plateaus.

1. *Old forest steppe*

The 'Túla-zaséki' and the forests on the Volga in Gov. Kazan are good examples. The former are part of the oak forests which at one time extended continuously through the Govs. of Kalúga, Túla, Tambóv and Kazan. It was called 'Zaséki' because it protected the arable land of the Russian peasants against the incursions of the steppe nomads. Trees were felled and built into an impenetrable wall to stop the advance of these invaders. It is noteworthy that Vysótski discovered relics of steppe chernozëm in the podsolised forest soil of the Túla-zaséki, together with bones of steppe rodents (probably *Spalax microphthalmus* Güld.) filled with the dark chernozëm soil, and here and there unleached calcium carbonate.

The oak forest, like the spruce forest, occurs on clay soils, but the oak does not tolerate complete podsolisation and requires greater warmth than the spruce. It is true that the oak penetrates far to the north in the coniferous forest region, but there it reaches furthest in the river valleys where the soil is warmer and less podsolised, giving way to the spruce on the elevated clay soils (Morózov).

2. *Existing forest steppe*

In this zone the forests penetrate far to the south along the river valleys forming 'gallery woods' along the high and steep river banks. The accompanying diagram (Fig. 15.1), from Tanfil'ev and Morózov, is applicable to a very large part of the steppe region between the rivers Dnieper and Volga. It will be seen that the left bank of the river (left of the diagram) is higher and steeper. The subsoil is clayey (loess-like and moraine clay), but here and there (not shown in the diagram) are abundant chalk outcrops. The right bank of the river passes into a raised terrace of fluvio-glacial sand. This sand, which was exposed to wind action and assumed the character of dune, was later occupied by pine forest.

The oak forest occupies the higher and drier portions of the right bank with well-grown trees, and also the steep left bank, passing some distance on to the plateau beyond. The characteristic changes of the soils are also shown in the figure. Above, on the plateau, which the oak forest did not reach so early, the chernozëm soil still exists, though it is degraded; passing thence into the river valley one meets first dark-grey forest soil, and then light-grey podsolised forest soil. In swampy places alder forest occurs. Further from the river, where sand begins to give place to clay, there is a peculiar mixed forest with an upper storey of pine and a lower storey of oak. Here and there, in the angle enclosed by the confluence of two considerable rivers and where the courses of the two rivers approach one another, the oak forest occupies extensive areas of the plateau. Such are the upland oak forests, of which the famous Tellermann's and Shípov's woods (Gov. Vorónezh), used by Peter the Great for the building of ships for the Sea of Azov, are examples.

The communities of these upland oak forests have a complicated structure, in correspondence with the soil changes, as shown in Tables 15.1 and 15.2.

We may conclude from these tables that the forest developed on the most favourable soil has the most complex structure and the best-grown trees, and that on the least favourable soil (solonéts) the trees are much more numerous in individuals and there is no

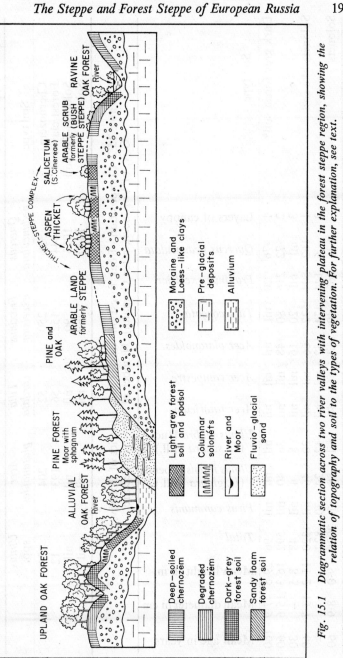

Fig. 15.1   *Diagrammatic section across two river valleys with intervening plateau in the forest steppe region, showing the relation of topography and soil to the types of vegetation. For further explanation, see text.*

Table 15.1 Number of stems (about eighty years old) per hectare in Shipov's wood, in relation to different soils (Stepánov)

Soil type	Quercus pedunculata	Fraxinus excelsior	Tilia cordata	Acer platanoides	Ulmus spp.
Chernozëm	318	250	52	204	24
Forest clay soil	444	152	56	193	12
Solonéts-like soil	1023	90	nil	nil	nil
Solonéts	1440	nil	nil	nil	nil

Table 15.2 Composition of oak forest on various soils (district of Bobróv, Gov. of Vorónezh, after Gumann). Number of stems per hectare

Soil type	Layers of canopy	Quercus pedunculata	Fraxinus excelsior	Tilia cordata	Acer platanoides	Acer campestre	Acer tataricum	Ulmus glabra Huds. (U. scabra Mill.)	Ulmus nitens Moench (U. glabra Mill.)	Pirus communis	Total	Mean height in m.	Mean diameter in cm	Mean age in years
Dark grey forest soil*	1	76	20	nil	nil	nil	nil	nil	nil	nil	96	29·5	77·9	160
	2	72	216	80	148	nil	nil	36	40	nil	592	16·3	15·6	60
	3	nil	nil	20	16	nil	nil	8	4	nil	48	8·5	–	30
Solonéts-like soil	1	162	504	522	66	nil	nil	20	46	nil	1320	11·2	13·7	60
	2	6	144	474	132	nil	nil	12	24	30	822	–	–	–
Solonéts	2	836	968	40	nil	56	48	nil	88	nil	2036	9·0	8·9	60

* Besides the three layers of tree canopy there are two lower storeys of vegetation: underwood and herb-layer.

one species dominant. The soil is occupied with difficulty and only as it is gradually changed. The trees are badly grown, much smaller in height and diameter, and the canopy is not differentiated into layers.

The following characteristics of these upland oak woods may be cited:

1. Since the oak unfolds its leaves very late, the lower storeys of the forest are well illuminated up to late in the spring, and consequently there is an abundant development of early flowering woodland herbs which give the forest the appearance of a flower garden.

2. In the summer there are few flowers and the herbaceous layer has the character of shade vegetation with leaf mosaics, etc.

3. There is no continuous moss layer owing to the masses of fallen leaves and the relatively thick herb layer. Mosses are confined to the lower portions of the tree trunks and to steep bare places on the edges of ravines, etc.

4. Myrmecochory is strongly developed among the herbs.

While the coniferous forests show a combination of tree growth with moss and lichen tundra, the deciduous forests show a similar combination with herbaceous communities. The steppe is dominated by herbs and though the herb communities of the forest differ both floristically and ecologically from those of the steppe, the steppe vegetation is in evidence in forest clearings, along tracks and wherever the tree canopy is not too close.

3. *Ravine forest*

This is practically confined to the bottoms and sides of ravines which intersect the plateaux. Besides the plants of the upland oak forest the following occur, especially in the damp ravines:

Athyrium filix-foemina	*Actaea spicata*
Cystopteris fragilis	*Geranium robertianum*
Dryopteris filix-mas	*Paris quadrifolia*
D. spinulosa	

Towards the south-east the oak penetrates a little even into the semi-desert zone, but here it is strictly confined to deep ravines with abundant freshwater springs, as on the Ergeni hills not far from Sarepta, which formed the old margin of the Caspian Sea. Here the oak is accompanied by very few representatives of the characteristic shade flora.

The oak in Russia is the representative of the central European deciduous forest, and correspondingly it grows best, according to Morózov, in the south-west and in Polés'e. In the east it suffers much from spring frosts. It has here a race which flowers and unfolds its leaves two to three weeks later in spring. Perhaps this can penetrate further east than the ordinary form. Near its northern and eastern limits the oak gradually loses its typical associates, so that the structure of the forest is modified. Thus in Gov. Vorónezh ash is freely mixed with oak, but further east the ash soon disappears.

According to topographic and edaphic conditions the oak forest may be divided into several types. Thus in each of the three regions, (*a*) west of the Dnieper, (*b*) central region, (*c*) east of the Volga, Vysótski divides it into the following types:

1. The dry side towards the steppe.
2. Massive central oak forest.
3. The moist side (relict oak forest where it gives way to the spruce).

Further details cannot be given here.

On the edge of the oak forest towards the steppe there is a fringe of trees and shrubs which tolerate dryness:

Acer tataricum	*Pirus communis*	*Rhamnus catharticus*
Crataegus	*P. malus*	*Ulmus nitens* Moench
monogyna		(*U. glabra* Mill.)

and of characteristic steppe shrubs:

Amygdalus nana	*Cytisus austriacus*	*Prunus fruticosa*
Caragana frutex	*C. ruthenicus*	*P. spinosa*
Spiraea crenifolia		

These trees and shrubs often also form independent communities of a peculiar dwarf wood or scrub, which is the first pioneer of woody vegetation on the steppe (and sometimes called 'Bush steppe'). On the drier and more exposed places this scrub consists mainly of *Amygdalus nana* and *Caragana frutex*, and transitions may be observed to the taller scrub of *Prunus spinosa* and *P. fruticosa* and to the edge of the steppe woodland. The larger shrubs form veritable 'gardens', many hectares in extent, and are exploited by the inhabitants in Govs. Tambóv and Saratov. Associated with them are tall herbs, grasses and dicotyledons, the latter decorating such places with numerous flowers. Among these are:

Agropyrum	*Lavatera thuringiaca*	*Origanum vulgare*
glaucum	*Libanotis montana*	*Veronica teucrium*
Bromus inermis	*Nepeta nuda*	
Chrysanthemum		
corymbosum		

and many others.

On the Ergeni the oak is already rare, but in valleys and depressions of the slopes little woods still occur with *Ulmus nitens* Moench (*U. glabra* Mill.) predominant and many of the shrubs and tall herbs mentioned above. Where the depressions are shallow these little woods peter out, and often pass over into societies of *Spiraea hypericifolia*, which give way, where the depressions are shallower still, to the regular herbaceous vegetation of the steppe. On the east of the lower Volga the oak is rare in the last islands of woodland and is confined to strongly leached clay soils, but the communities of dwarf wood penetrate far into the semi-desert, and here *Spiraea hypericifolia* is characteristic and widely distributed.

GRASS STEPPE – SEMI-DESERT – DESERT

As one passes to the south, and especially to the south-east, the climate becomes drier and the summers hotter, and there is a gradual transition from grass steppe with its fertile 'black earth' soil (chernozëm) to the desert of Turkestan with 'grey earth' (serozëm) soil. The changes of the chief kinds of vegetation as we pass through the various soil-type zones are as follows.

The oak woods which are present on the northern chernozëm soils disappear on the leached chernozëm of the northern grass steppe region. Perennial and biennial dicotyledons, on the other hand, became abundant on the leached chernozëm, decrease on the deep-soiled and middle chernozëm types and rise again on the southern chernozëm, to reach a secondary maximum on the chestnut soils of the semi-desert. With drier conditions still they decrease on the burozëm (brown earth) soils and are absent in the true desert (serozëm) soil. Inversely the tussock steppe grasses, especially *Stipa* and *Festuca sulcata*, increase from the leached to the middle chernozëm and decrease again on the southern chernozëm and chestnut soils. Like the dicotyledons they do not reach beyond the burozëm soils. The xerophilous undershrubs with whitish felted leaves first appear on the southern chernozëm and increase steadily to the burozëm soils; they are still represented in the desert. The spring

ephemerals exist in small numbers on all the chernozëm soils and
rise proportionally through the semi-desert to the desert soils. Finally
the lower cryptogamic plants of the soil surface are but slightly
represented on the chernozëm soils, rise to a maximum on those of
the semi-desert and fall again in the true desert.

In the course of this series the leaching of the soil in general
diminishes from the more northern chernozëm to the serozëm. In
the deep-soiled chernozëm 10 per cent. HCl effervesces only at a
considerable depth, e.g. of more than a metre, but in the serozëm as a
rule on the surface. According to Tumin the central region of deep-
soiled chernozëm presents the most favourable conditions for the
accumulation of humus, because on the one hand the soil is not so
strongly leached as to impoverish it seriously in valuable nutritive
material and on the other hand the effect of drought is not too
severe. But nowhere in the series is there any accumulation at the
surface of easily soluble sodium salts (NaCl and Na_2SO_4). In the
light-chestnut and brown-earth soils it is true that there is considerable accumulation of sodium sulphate, but this only begins at a depth
of about 1·5 m.

MEADOW STEPPE IN THE SUBZONE OF DEEP-SOILED CHERNOZËM

At the height of its development at the end of May and the beginning
of June the closed vegetation has the appearance of a flowering
meadow. For instance, in the district of Vorónezh on 11 June 1915,
such a steppe showed innumerable bluish-lilac flowers of *Campanula
simplex*, white heads of *Trifolium montanum* and fragrant cream-
coloured inflorescences of *Filipendula hexapetala*, with many other
bright conspicuous blossoms. *Bromus erectus* was present in great
masses with many inflorescences. But the basis of the vegetation was
made up of the sod-forming steppe grasses *Festuca sulcata* and
Koeleria gracilis. *Stipa joannis*, *S. stenophylla* and *S. capillata* were
also present, but they were more or less scattered and did not suppress other species. The steppes described by Sprygin in Gov. Pénza,
also on deep-soiled chernozëm, have a similar character. Besides the
dicotyledonous herbs mentioned above, *Galium verum* and *Hypo-
chaeris maculata* may be added as well as other species which appear
where the soil is moister, e.g. *Chrysanthemum leucanthemum*,
Polygonum bistorta and *Sanguisorba officinalis* (Plate 11(*b*)). This
type of steppe may be briefly characterised as follows.

1. The turf-forming grasses form the basis but do not suppress numerous dicotyledonous herbs, which mostly form basal rosettes, but also raise assimilating leaves relatively high above the soil surface. The dominance of the sod-forming grasses in the northern variants is, according to Tumin, to be related to the relatively low fertility of the soil, and both the sod-forming habit and the basal rosettes of the dicotyledons may also be correlated with the cutting of hay which is the rule in steppes of this type.

2. Spring ephemerals play but an insignificant part below the carpet of perennial herbs.

3. Lower plants, such as mosses, are conspicuous only in the northern variants of this vegetation.

The regular mowing of this type of steppe cannot fail to exercise an influence on the vegetation, but little has yet been done towards the solution of this problem. It is highly probable that species of *Stipa* and especially *S. stenophylla* were originally more widely distributed on these steppes, particularly on their southern portions, and that they approximated to the types of *Stipa* steppe. The cutting of hay would without doubt hinder the reproduction of this species from seed, since it ripens its fruits later than the ordinary time of hay harvest.

STIPA TUSSOCK STEPPES

The deep-soiled chernozëm region and its vegetation passes over gradually towards the south and south-east into the ordinary or middle chernozëm region, and the meadow steppe also passes over into steppe characterised by the larger *Stipa* tussocks. It is not, however, by any means certain that this soil transition corresponds exactly with that of the vegetation. *S. stenophylla* and *S. capillata* are dominant on the *Stipa* tussock steppe, and the smaller turf-forming grasses such as *Festuca sulcata* and *Koeleria gracilis*, though widely distributed, play a subordinate part. The *Stipa* roots branch profusely in the upper layer of soil and use most of the available water, thus restricting the dicotyledonous herbs, which, while common enough, are only scattered and do not form the gay flower garden of the meadow steppe. Such deep-rooting species as *Medicago falcata*, however, hold their own well with the stipas, tapping a deeper layer of soil and remaining green and fresh late in the summer when the grasses are more or less burnt up (Gordyágin). The lower plants are not much in evidence (*Tortula ruralis* and *Nostoc commune* appear in consequence of cattle grazing) and the spring ephemerals, such as *Hyacinthus leucophaeus* and *Poa bulbosa* var. *vivipara*, have but a minor importance.

These *Stipa* steppes are ordinarily used for cattle grazing and it is not clear how far this use has an effect on their constitution. Nor is it clear from the literature whether *Stipa stenophylla* is most widely distributed on the ordinary or on the deep-soiled chernozëm.

Small tufted Stipa *steppes*

Here again the *Stipa* tussocks gradually lose their dominance, but not as in the meadow steppes owing to increase of moisture, rather indeed because of increased dryness. *Stipa tirsa* and *S. lessingiana* with small tussocks are the species here and *Festuca sulcata* again acquires more importance. Dicotyledonous herbs have room for their root development but their leaves are borne closer to the soil surface. So we gradually pass from the grass steppe to the semi-desert vegetation.

FACTORS HINDERING THE SPREAD OF FOREST SOUTH-EASTWARD

The main factors militating against the occupation of the extensive southern and south-eastern steppes by forest, according to the views of various Russian authors, are as follows:

1. Lack of moisture in conjunction with the fine-grained steppe soil. The precipitation of the steppe climate is markedly lower than that of the forest climate, more of the water is lost to the soil through immediate evaporation as a result of the steppe winds, penetration to the deeper layers of the soil is hindered by the fine-grained nature of the surface, and the water which does penetrate cannot be so completely utilised by the root systems.

2. The strong steppe winds also have of course a direct effect on transpiration. Wind is in general the greatest enemy of trees, and not only the dry south-east winds of summer (often accompanied by the so-called 'dry fog', Russian *mgla*) but probably also the winter winds have a great effect in preventing the spread of trees over the steppes just as they do over the northern tundra, since Gordyágin and Ivanov have shown that trees, even in the leafless state, may lose considerable quantities of water.

3. The presence of large quantities of easily soluble salts harmful

to certain of the forest vegetation. But a clear picture of these effects is only seen where the deciduous forest reaches the solonéts soils, and even here a doubt remains as to whether the peculiar physical character of the soil does not affect the trees unfavourably. On the edge of these soils the sufferings of the trees are clear enough. They are dwarfed, of irregular growth, and the top, or even the whole tree, may perish: lichens settle in great numbers on the bark. The forest passes over the solonéts soils and cannot colonise them, at any rate without centuries of preparation. Patches of solonéts, still surrounded by crippled trees, are yet to be found in the forest. But where forest borders on pure steppe (i.e. with chernozëm soil) such pathological phenomena are not to be found.

4. Competition with the steppe grasses is probably also a factor. If acorns or the fruits or seeds of other trees are strewed among the grasses of a typical *Stipa* steppe (which as already described form massive tussocks and exploit very thoroughly the water of the surface layer of soil), the seedlings abort. But if the grasses were completely extirpated without altering the soil itself, the development of tree seedlings would probably take place.

It can be seen that the distribution of forest in the drier part of the steppe zone is often localised in places which from various causes are physiologically moister. The ravine woods and 'gallery' woods are cases in point. And the farther one goes to the south-east in the dry steppe zone and then into the semi-desert and desert, the more clearly and closely forest is confined to damper and more leached soils. In the so-called 'forest–steppe' zone, however (i.e. in the transition region between forest and steppe), the forest often occupies areas which are apparently identical in conditions of soil and land relief with other areas which are occupied by steppe vegetation. And clear evidence of the advance of forest on steppe is not lacking. Thus forest invades the plateaux from the river valleys and can to a certain extent prepare the conditions for its own successful advance. Thus the wind has less effect on a close-set phalanx of trees than on isolated individuals, snow collects along the forest edge, and on its melting the snow water moistens and leaches the soil, leading to podsol formation and less favourable conditions for the steppe grasses, which are also shaded by the advancing trees. In the northern parts of the forest–steppe zone this advance is very marked. Burial mounds, which were undoubtedly made in open steppe, are now covered with trees.

Nevertheless in this northern region steppe often appears to be protected from forest invasion, and this has been attributed to fire, which might originate from lightning. Also the inhabitants of these regions have been acquainted with the use of fire since palaeolithic times. Such fires due to human agency may have been partly accidental, but it is also possible that the steppe was deliberately burned in order to destroy the litter of dead leaves and stems which hinder new growth of the steppe plants. Again in certain historical periods fire was used by the Moscow kingdom as a means of protection against invading Tartars. The grass was burned over immense stretches so that the Tartars should find no fodder for their horses. Thus for instance in 1571 Prince Vorotýnski and his associates, by order of the Tsar, decided to set fire to the steppes at the time of the first autumn frosts when the grass was dry and before the snow fell, in clear weather and when the wind was blowing from the border towns towards the steppe. The Cossack villagers were ordered to burn the steppe in this way from the sources of the Vorona as far as the Dniester and Desná.

We cannot, however, regard the problem of the causes of the hindrance of the advance of forest on steppe as definitely solved. Many possible factors have been indicated and indeed every conceivable explanation has been brought forward, just as in the case of the American prairies. The proper weight and relative importance of the single factors have not been settled. Especially, little attention has been devoted to the different species of trees in this connection. The causes of the failure of the pine, for instance, to spread on to the steppe may be quite different from those hindering the advance of oak or aspen. Certainly trees exist or could be produced, well adapted to occupy steppe. Even in desert such trees as *Arthrophytum* are able to invade. But the steppes are geologically recent and the available trees are few and are adapted on the whole to a damp climate. The characters of the individual species available must always be taken into account.

BIBLIOGRAPHY

Since it is impossible to give an even approximately complete list of the vast literature relating to the theme of my sketch of Russian vegetation, I cite only a few general works which mostly deal with extensive regions and contain many references, together with a few (marked *) written in German.

ALEKHIN, V. V., *The Plant Covering of the Steppes in the Central Chernozëm Region* (Russian) (Vorónezh, 1925).

BUSCH, N. A. *Phytogeographical Sketch of European Russia* (Russian) (Academy of Sciences, Petrograd, 1923).

DOKTUROVSKI, V. S., *Bogs and Peat-moors, their Development and Structure* (Russian) (Moscow, 1922).

*—— *Über die Stratigraphie der russischen Torfmoore* (Stockholm, 1925).

FOMIN, A. V., *Short Sketch of the Phytogeographical Regions of the Ukraine* (Russian) (Kiev, 1925).

*GLINKA, K. D., *Die Typen der Bodenbildung* (Berlin, 1914).

—— *The Soils of Russia and Adjacent Countries* (Russian) (Moscow–Petrograd, 1923).

GORDYAGIN, A. Y., *The Vegetation of the Tatar Republic* (Russian) (Kazan, 1922). (*Geographical Description of the Tatar Republic*, Part I.)

KARPINSKI, A., *Sketches of the Geological History of European Russia* (Russian) (Petrograd, 1915).

KELLER, B. A., *The Plant-world of the Russian Steppes, Semi-deserts and Deserts* (Russian) Part I (Vorónezh, 1923).

—— 'Die Vegetation auf den Salzböden der russischen Steppen, Halbwüsten und Wüsten (Versuch einer ökologischen Präliminäranalyse)', *Zeitschrift für Botanik*, XVIII (1925–6).

*—— 'Die Grassteppen im Gouvernement Woronesh, Russland', in Karsten und Schenk, *Vegetationsbilder*, Reihe 17, Heft 2.

*KOPPEN, FR., *Geographische Verbreitung der Holzgewächse des Europäischen Russlands und des Kaukasus*, Teile 2 (1888–9).

KORZHINSKI, S. T., 'The Vegetation of Russia' (Russian) in Brockhaus and Efron, *Encycloped. Lexicon*, XXVII (Russland) (Petersburg, 1899).

KOZO–POLYANSKI, B. M., *The Alpine Rose of the Central Chernozem Region of Russia: A Contribution to the Knowledge of the Relict-flora of the Central Russian Uplands* (Russian) (Vorónezh, 1926).

LITVINOV, D. Y., 'Geobotanical Notes on the Flora of European Russia' (Russian), *Bulletin of the Society of Naturalists of Moscow* (1890).

—— *On the Relict Characters of the Flora of the Rock Slopes in European Russia* (Russian), Publications of the Botanical Museum of the Academy of Sciences, I (Moscow, 1902).

MOROZOV, G. F., *Forest Lore* (Russian) (1925).

—— *Forest Types in Relation to their Significance for Forestry* (Russian) (1917).

NOVOPOKROVSKI, I. V., *The Vegetation of the Don Region* (Russian) (Novocherkassk, 1921).

—— *The Vegetation of the North Caucasian Region* (Russian) (Rostov on the Don, 1925).

PACZOSKI, I. K., 'Leading Characteristics of the Development of the Flora of South-western Russia' (Russian), *Annals of the New Russian Society of Naturalists*, XXXIV (Odessa, 1911).

—— *Description of the Vegetation of the Government of Kherson:* I. *The Woods* (Russian) (Kherson, 1915); II. *The Steppes* (Kherson, 1917).

*POHLE, R., 'Pflanzengeographische Studien über die Halbinsel Kanin', *Acta Horti Petropolitani*, XXI, Lief. 1 (1903).

PRASOLOV, L. I., *The Soil Regions of European Russia* (Russian) (Petersburg, 1922).

SUKACHEV, V. N., *Swamps, their Formation, Development and Characteristics* (Russian) (1923).

—— *Plant Communities* (1926).

TALIEV, V. I., *Vegetation of the Cretaceous Exposures of South Russia* (Russian), Publications of the Kharkov Society of Naturalists (Kharkov) Part I (1904); Part 2 (1905); Supplem. (1907).

TANFIL'EV, G. I., *The Geography of Russia, the Ukraine and the Adjacent Countries to the West* (Russian) Part 2, Sect. 1, 'The Relief of European Russia and the Caucasus' (1922); Part 2, Sect. 3, 'Terrestrial Magnetism, Climate, Rivers, Lakes' (1924) (Russian).

—— *Outlines of the Vegetation of Russia* (Russian) (1903).

—— *Forest Limits in Arctic Russia* (Russian) (Odessa, 1911).

—— *Forest Limits in Southern Russia* (Russian) (Petersburg, 1894).

VYSOTSKI, G. N., 'Ergeni' (Russian), *Bulletin of Applied Botany*, VIII (Petrograd, 1915).

16 The Parkland or Grove Belt of Alberta

E. H. MOSS

From 'The vegetation of Alberta: IV. The poplar association', *J. Ecol.*, xx (1932) 382, 401–7.

GENERAL DESCRIPTION

THE terms 'Parkland' and 'Grove Belt' are applied, in this paper, to a transition or tension belt which lies between the Poplar Area and the Prairie (Fig. 16.1). This belt consists of groves of aspens and patches of prairie grassland, more or less uniformly intermixed. In the central part of the belt, these two types of vegetation dominate practically equal areas, the aspen community occupying the more moist and more sheltered situations, the prairie occurring in the drier and more exposed places. Such typical Parkland is illustrated in Plate 12(*b*), which shows a knoll viewed from the east with aspen vegetation on the north slope and prairie on the south slope. In areas of this kind the aspen consociation also occurs on north-east and sometimes on east slopes, while prairie usually occupies west slopes.

Near the southern limit of the Parkland, prairie vegetation occupies a large proportion of the region, wooded vegetation being mainly confined to the margins of sloughs, banks of streams and north-facing slopes of deep ravines and river valleys. In these situations, the wooded vegetation may be dominated by poplar, but very often the only trees present are small willows, prevailing species being *Salix petiolaris*, *S. discolor*, *S. bebbiana* and *S. interior* Roulee. Towards the northern and western limits of the Parkland, on the other hand, the aspen consociation comprises the bulk of the natural vegetation and prairie vegetation is generally restricted to the steeper and drier southern exposures. In passing, it may be pointed out that prairie vegetation is of occasional occurrence in the Poplar Area, for example, on very dry south-facing slopes of river valleys; also, prairie vegetation tends to invade parts of the Poplar Area that have been drained, frequently burned and kept clear of trees for a period of years, for example, roadsides.

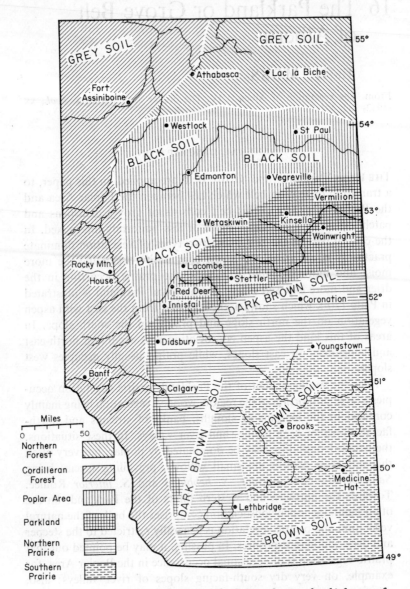

*Fig. 16.1   Map of southern and central Alberta, showing the chief types of
vegetation and soils*

THE PRAIRIE VEGETATION

The Prairie is a complex formation dominated by numerous grasses which are variously intermixed or grouped. In Alberta, the Prairie consists of two rather well-defined communities which we name the Northern Prairie and the Southern Prairie. Whether these communities should be regarded as associations is a question requiring further investigation. Although there is a broad transition belt between the Northern and Southern Prairies, and although there occur 'islands' or 'outliers' of each, one within the other, yet we may represent, in a general way, the areas occupied by them, as in Fig. 16.1. It will be noted that the Southern Prairie coincides closely with the Brown Soil belt, while the Northern Prairie occupies the greater part of the Dark Brown Soil belt.

The dominant and certain subsidiary species of the Prairie are shown in Fig 16.2, in which the black bands represent our conception of the occurrence and relative abundance of the different forms. It should be pointed out that we are here considering only typical prairie vegetation, such as forms the natural cover of a very large proportion of the region, and are ignoring the vegetation of various special habitats occurring within the prairie, for example, sandy areas, alkali flats, coulées and other depressions. The leading grasses of the Southern Prairie vegetation are *Bouteloua gracilis* (H.B.K.) Lag., *Stipa comata* Trin. and Rup., *Agropyron smithii* Rydb. and *Koeleria gracilis* Pers. (the latter may perhaps be more properly regarded as a member of the Northern Prairie community); while grasses of secondary importance include *Stipa viridula* Trin., *Poa* spp., *Agropyron dasystachyum* (Hook.) Scribn. and certain other species of *Agropyron*.

Clarke [1], in a paper that came to hand after the present paper was accepted for publication, gives the most reliable account yet available of the vegetation of the Southern Prairie or Short Grass Plains of Western Canada. Clarke regards *Agropyron smithii* as a somewhat less important constituent and *Poa* spp. as considerably more important members of the Southern Prairie than is reported in the present paper.

The leading species of the Northern Prairie region include several of those that are prominent in the Southern Prairie, notably *Koeleria gracilis*, *Stipa* spp., *Agropyron* spp. and *Bouteloua gracilis*, and, in addition the following: *Festuca scabrella* Torr., *Avena hookeri*

Scribn., *Agropyron tenerum* Vasey and *A. richardsoni* Schrad. and *Poa interior* Rydb. Grasses of secondary importance are *Muhlenbergia squarrosa* (Trin.) Rydb., *Danthonia intermedia* Vasey, *Agrostis hyemalis* (Walt.) B.S.P., *Bromus porteri* (Coult.) Nash, and several species of *Poa*, of which *P. pratensis* L. and *P. palustris* L. have become dominant in many places, especially in Parkland areas where the natural vegetation has been more or less destroyed, for example, on roadsides.

As indicated in Fig. 16.2, the chief grasses of the Northern Prairie vegetation may be arranged in approximate order of importance as

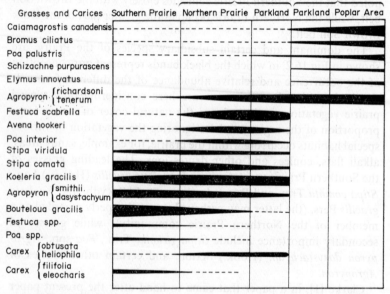

Fig. 16.2 A graphical representation of the occurrence and relative abundance of leading grasses and sedges in various regions and communities

follows; *Koeleria gracilis, Festuca scabrella, Stipa comata, Agropyron tenerum, A. richardsoni, A. dasystachyum, A. smithii, Avena hookeri, Stipa viridula, Bouteloua gracilis* and *Poa interior.* There is, however, some doubt whether certain of these grasses (namely, *Bouteloua gracilis, Agropyron smithii, A. dasystachyum,* and perhaps also *Stipa* spp.) are to be regarded as important constituents of the Northern Prairie community; they are usually abundant or dominant in the more xeric situations of the Northern Prairie region and may perhaps be properly thought of as 'islands' of the Southern Prairie community in the Northern Prairie region. Associated with the grasses in the Northern Prairie are certain small sedges, of which *Carex obtusata* Lilj. and *C. heliophila* Mackenzie are often abundant. Other common and characteristic species of the Northern Prairie community are as follows: *Cerastium arvense* L., *Pulsatilla ludoviciana* (Nutt.) Heller, *Sieversia ciliata* (Pursh.) G. Don., *Comandra pallida* DC., *Selaginella densa* Rydb., *Symphoricarpos occidentalis* Hook., *Gaillardia aristata* Pursh., *Achillea lanulosa* Nutt., and species of *Aster, Artemesia, Antennaria, Solidago, Helianthus, Grindelia, Chrysopsis, Rosa, Potentilla, Fragaria, Androsace, Thermopsis, Astragalus, Oxytropis, Petalostemon, Pentstemon, Heuchera, Linum, Arabis, Anemone* and *Zygadenus*.

The relationship between Northern Prairie and Southern Prairie vegetation may be further considered by reference to Fig. 16.2. It will be observed that as one passes from one type of prairie to the other, practically all of the dominant grasses show either a marked decrease or increase in abundance. A similar relationship holds for various subsidiary grasses, for certain carices, and also for numerous species not included in Fig. 16.2. These features, taken in conjunction with certain facts regarding the habitat relationships and grouping of the leading species (referred to in the preceding paragraph), appear to provide a sound basis for the recognition of Northern and Southern Prairie communities. Of these, it is the Northern Prairie that usually comes into immediate conflict with the poplar association and it is this community therefore which is of particular interest in the present discussion. By way of summary, we may state tentatively that the Northern Prairie community is composed of the dominant grasses, *Koeleria gracilis, Festuca scabrella, Agropyron richardsoni, A. tenerum* and *Avena hookeri*, and numerous secondary species, including various grasses, carices and other common and characteristic species already noted.

ECOTONE BETWEEN PRAIRIE AND POPLAR VEGETATION

The Parkland has already been described as a broad transition or tension belt consisting of patches of poplar and prairie vegetation. This transition belt contains within it thousands of true tension lines or ecotones, these occurring wherever poplar and prairie communities meet. Photographic illustrations of tension lines may be referred to here. In Plate 12(*b*) a narrow ecotone extends along the top of the elevation between the prairie vegetation on the south slope and the aspen consociation on the north slope, and consists of representatives of the two competing communities. Leading forms include various grasses and such shrubs as *Symphoricarpos*, *Rosa*, *Prunus*, *Elaeagnus argentea* and *Shepherdia canadensis*. The ecotone also extends along the east face and again along the west face of the elevation and in these situations is usually a rather broad band. In Plate 12(*a*) is shown a typical ecotone – a narrow belt, dominated by the 'wolfberry', *Symphoricarpos occidentalis*, bordering the aspen grove. The wolfberry extends into the prairie, smothering the grass, and is followed by the aspen, shoots of which come up in the wolfberry belt, eventually becoming dominant there. In many situations the aspen vegetation thus tends to invade and to succeed the prairie vegetation; and the border of the aspen grove then becomes composed of trees that are considerably younger than those of the main body of the stand. Determinations of the ages of trees in the border of one such grove indicated that the rate of advance of the trees during the last twelve years has been about 15 cm. per year. It should be pointed out here that the occurrence of younger trees on the border of a grove is not always to be explained as above, but rather may be the consequence of one or more burnings of the marginal part of the grove, the younger trees having sprung from the underground parts of the burned trees. A situation of this kind is usually recognisable by the fact that the trees of the greater part of the bordering belt are of the same age; whereas, in a belt formed as a result of invasion of the prairie, the trees show a progressive increase in age from the periphery inwards.

This marked tendency for the aspen consociation to invade and replace prairie vegetation is convincingly illustrated in many parts of the Parkland; and, in the absence of burning, succession in this direction would soon produce considerable extension of woodland. That burning has been effective in stemming the advance of trees is clearly shown by certain areas investigated. On the other hand, where fires have not occurred or have been infrequent, for several years, as is usually true of regions that have been for some time settled and in part under cultivation, the trees have made marked advances upon the prairie. Early settlers of the Parkland region assert that groves are now more numerous and more extensive than they were a number of years ago.

There seems to be good reason therefore to conclude that fire has been a very important factor in preventing expansion of the aspen consociation in the Parkland, and probably too in limiting the southern extension of the Parkland area. During past centuries the frequency of burning has doubtless fluctuated very considerably. In very dry periods frequent fires no doubt invaded the region, burning off aspen groves and possibly killing some of them as a whole or in part. This appears to have been the situation in the Parkland and also in the southern part of the Poplar Area about 1890 when many of the early settlers arrived, and may be the foundation of the popular belief that various regions now well wooded were at one time prairie. During the subsequent wet period, numerous aspen groves appeared in the region, possibly from seed, but more likely, in the main, from roots that had survived the dry period and the ravages of frequent fires.

While fire is considered to be the chief factor responsible for checking the invasion of prairie by the aspen, certain other factors are at times quite effective. The snowshoe rabbit, *Lepus americanus phaeonotus* Allen, by girdling young aspens, exerts a marked effect, at least during periods when this species is at a maximum. Bird [2] states that 'the aspen would advance much more quickly on to the prairie, if it were not for the rabbits'. The buffalo, *Bison bison* L., doubtless an important factor in the past, ceased to be effective some fifty years ago, when the last of the great wild herds were slaughtered. In recent years, man has interfered greatly with the development and extension of aspen groves by cutting trees, by setting fires, by cultivating areas adjoining groves and by draining certain areas. On the whole, however, and excluding from consideration regions that are now largely under cultivation, man has tended to shift the balance in favour of aspen vegetation, because he has prevented the frequent occurrence of widespread burning.

The striking correlation, already noted, between the occurrence of the two types of Parkland vegetation and the topography of the region appears to be a consequence of three interrelated circumstances, namely:

1. Hill tops and south-facing slopes, being generally hot, dry situations, are suited to the development of prairie vegetation but do not favour the invasion and growth of trees; whereas depressions and north-facing slopes, being much less xeric, favour the growth of trees and the consequent exclusion of prairie grasses.

2. Burning is common in the drier situations but much less frequent in the more moist habitats, with the result that invasion of prairie vegetation by trees is counteracted; therefore, restriction of the two vegetational types to their characteristic habitats tends to be maintained.

3. Ecesis of the aspen probably occurs very rarely in the drier situations, but not infrequently in the more moist places. This subject is considered at length in the following paragraphs.

At some time in the past, whether in comparatively recent years or many centuries ago, the thousands of isolated aspen groves of the Parkland must have been initiated by seed. The question of the time factor involved here may be elucidated in the future by a critical study of the soils on north and south slopes. At present we are inclined to believe that groves of trees have been abundant in the greater part of the Parkland for many years but not for many centuries. Even the Poplar Area seems not to have been dominated by trees for a very long period of time. For the soils of most of the Parkland and Poplar Areas are of the grassland rather than the wooded type. Furthermore, with the exception of the black soils in the extreme northern and western parts of the Poplar Area, these soils show only a slight transformation of the kind expected in grassland soils that had been wooded for many centuries.

Establishment of the aspen in grassland appears to be dependent upon the following conditions: (1) The transfer of viable seed; how far aspen seed may be transported in a viable condition is problematical. (2) Local climatic and edaphic conditions favourable to the germination of the seed and the establishment of seedlings, conditions that may rarely obtain, at least while the seed is still viable. Although no measurements have been made, there is undoubtedly a great difference in this respect between north and south slopes, the latter subject to direct insolation, being much hotter and drier. (3) An open prairie community, such as is likely to occur on steeper slopes, as a consequence of erosion. Doubtless, there are occasional years in which all of these conditions are fulfilled and when the aspen may, therefore, become locally established in the prairie community, especially in depressions and on north and north-east slopes.

Another way in which aspens may become established in the prairie will now be considered. Within the prairie community of the Parkland there occur numerous stands of the wolfberry, *Symphoricarpos occidentalis*. In the southern part of the Parkland these stands commonly occupy shallow depressions and northern exposures, while farther north they occur in drier habitats, often on hillocks. On lighter soils, *Symphoricarpos* is usually accompanied by another shrub, the silverberry, *Elaeagnus argentea*. Not uncommonly, these shrubs occur as co-dominants, forming two quite distinct layers, the greyish silverberry projecting far above the deep green wolfberry. These communities, the first dominated by *Symphoricarpos*, the second by *Symphoricarpos* and *Elaeagnus*, may be regarded as societies or possibly as associes. Accompanying species include *Rosa* spp., *Rubus strigosus*, *Galium boreale*, *Lathyrus ochroleucus*, *Solidago* spp., *Anemone* spp., *Agropyron* spp. and other grasses. According to Bird [2], who has recently studied the eastern part of the Canadian Parkland, these shrub communities have become established in grassland as a consequence of the activities of certain animals; and our own observations point to the same conclusion. There is also considerable evidence that the shrub communities are, in some cases, succeeded by the aspen consociation, the aspen being initiated by seed. Bird claims that succession along these lines is largely due to the activities of the following animals: 'gophers' or ground squirrels, including Richardson's ground squirrel, *Citellus richardsoni* and the pocket gopher, *Thomomys talpoides rufescens*; the badger, *Taxidea taxus*, the chief predator of *Citellus*; and the pine grosbeak, *Pinicola enucleator*. According to Bird, gophers and badgers throw up mounds of earth that choke out the grass, thus enabling *Elaeagnus* and *Symphoricarpos* to gain a foothold, seeds of the latter shrub being distributed in large numbers by the pine grosbeak. In the comparatively loose earth and sheltered conditions of *Symphoricarpos* and *Symphoricarpos–Elaeagnus* stands, aspen seedlings may become established.

REFERENCES

[1] CLARKE, S. E. (1930) 'Pasture investigations on the short grass plains of Saskatchewan and Alberta', *Scientific Agriculture*, x, no. 11.
[2] BIRD, RALPH D. (1930) 'Biotic communities of the aspen parkland of Central Canada', *Ecology*, xi, no. 2.

17 Climatic Change or Cultural Interference? New Zealand in Moahunter Times

KENNETH B. CUMBERLAND

From 'Climatic change or cultural interference? New Zealand in moahunter times', in *Land and Livelihood: Geographical Essays in Honour of George Jobberns* (N.Z. Geographical Society, Christchurch, 1962) pp. 124–35.

THE USES OF FIRE

APART from the energy of human muscles, fire was the only power at the neolithic Polynesian's command. It was capable of wielding tremendous influence on his environment. In the tropical island homelands there had been little need for it: but in New Zealand, it seems very likely that the kindling, preservation and transportation of fire became of major concern. The choice of sites (as distinct from locations) of hearths, ovens and hutments was probably influenced by the availability of driftwood as much as by any other site factor. The moahunter used fire not only in his river-mouth settlements and around his permanent hearths, but carried it and kindled it on exploratory journeys inland. Through accident and inadvertence, if not by experiment or design, camp fires were left burning, or escaped into the vegetation around; and in an instant fire brought ecological disturbance to virgin plant cover – grassland, scrub, and forest margin – such as slowly operating climatic change might take many decades, even centuries, to accomplish.[1] There was little reason for avoiding fires or for putting them out once alight. Because of the very sparsity of his occupance the moahunter could do little harm. Indeed he may soon have found that advantage and reward were to be derived from what at first was accidental firing of the vegetation.[2]

[1] Professor Cumberland's main aim in writing this essay was to refute the views of archaeologists and foresters such as J. T. Holloway, who have held that the vegetation changes which took place some eight centuries ago in New Zealand were due to climatic deterioration. See John T. Holloway, 'Forests and climates in the South Island of New Zealand', *Trans. Roy. Soc. N.Z.*, LXXXII (2) (Sep 1954) 324–410 (Ed.).

[2] See G. Kuhnholtz-Lordat, *La Terre Incendée* (Nîmes, 1939); Walter Hough, 'Fire as an agent in human culture', *U.S. Nat. Mus. Bull.*, no. 139 (Washington, 1926); O. C. Stewart, 'Burning and natural vegetation in the United States', *Geogr. Rev.*, XLI 2 (Apr. 1951) 317–20; and 'Fire as the first

He may well have learned, as did the pastoral runholder in the 1850s, that accidental fires, sweeping across the plains of grass and damaging the forest margins, improved and extended the moa grazings, freshened up the grass and so attracted bigger numbers of hungry birds. As the moahunter apparently developed no weapons other than a small bird-spear for hunting the flightless birds, he is soon likely to have found fire an aid to the capture of his quarry. Not only would a raging tussock fire destroy eggs and kill and injure moas in its passage, and leave them scorched and dying on the plains; but, like the sheepdogs of later immigrants, it would drive them in flocks to places where they could be rounded up and despatched with facility – on the wet and boggy margins of swamps, on no-exit shingle spits and sandbars, and in the expansive dry beds of rivers between deep, anastomosing streams.

If the South Island's first settlers did accidentally discover such practical motives for incendiarism, it is likely that firing of the vegetation would become a regular and purposive practice. A repeated use of fire would have two significant results. It would aid climate, especially in the drier oscillation up to as late as A.D. 1200, to extend the range of the open grassland at the expense of trees and shrubs; and it would aid in reducing (and ultimately in extinguishing) the moa population, and in driving the fleeter, lighter species deep into the inland, grassy basins of the South Island and from the open grassland into the relative safety of the forest.[1]

As moa species and other avifauna including swan, giant rail, goose and eagle became extinct, and as others were remarkably reduced in numbers, perhaps some centuries after initial settlement, the moahunter's quarry had to be followed inland, and into the

---

great force employed by man', in Thomas, *Man's Role in Changing the Face of the Earth* (Chicago, 1957) pp. 115–33. Stewart believes 'that native peoples have rarely been careful in extinguishing campfires . . . [and] that hunting and gathering peoples from the time they acquired fire have allowed their fires to ignite the landscape because it did not occur to them to protect the vegetation from fire' (in Thomas, p. 118). To the moahunter the preservation of fire may have been of greater concern than putting it out.

[1] It is probably no accident that on present evidence the larger genera and species would appear to have survived longest in the damper forested southernmost areas of the South Island, that the small bush-dwelling moa, *Megalapteryx*, apparently survived west of Lake Te Anau into the seventeenth or eighteenth century and the age of metal tools, or that it was above the tree line in remote Fiordland that the flightless rail, *Notornis hochstetteri*, was rediscovered in 1948.

matai forests. Although the moahunter came to rely more and more on fish and shellfish, as is suggested by changing emphases in the sequence of culture layers in Murihiku,[1] the moa became more prized and more important as its numbers were reduced. Man had learned to use its skin for clothing, certain bones for making ornaments, fish-hooks, spearpoints, harpoon points, minnow shanks and awls, and eggs for pierced shell water vessels.

Seasonally, or permanently, some settlements were transferred inland, as for example in the Hawkesburn Valley, Central Otago.[2] In other cases there is the possibility of hunting camps being established inland from which carcases were rafted by water to long-established coastal settlements, as for example from the Mackenzie country to the Waitaki mouth.[3] When the moas which inhabited the driest inland grassy basins had been reduced in numbers, the hunter had to find his quarry in the bush. Although he used and worked timber (as the richness of his carpentry tools indicates), the moa-hunter never came to regard the forest as a basic and valuable re-source. His economy was based on the yield of the sea, lagoon and grassland. Like his European successor he found the forest an obstacle rather than an advantage. He allowed grass and scrub fires to push back the forest margin; and, when the resource upon which his economy now principally rested could only be procured in the forest, he doubtlessly utilised fire to destroy the moa's last natural protection.

Before the onset of a rather damper and cooler oscillation of climate after A.D. 1200, at a time when much of the podocarp forest east of the ranges – established originally in the post-glacial Climatic Optimum – survived in somewhat precarious ecological balance, the destruction of much of the forest by fire would not be difficult. By choosing aright the time for firing – towards the end of the period of nor'west winds which prevailed in the spring and early summer and at the approach of the wilting point of many undergrowth species –

[1] L. Lockerbie, 'From moa-hunter to classic Maori in southern New Zealand', in J. D. Freeman and W. R. Geddes, *Anthropology in the South Seas* (New Plymouth, 1959) pp. 75–110.

[2] Ibid., pp. 85–7.

[3] H. S. McCully, quoted in Roger Duff, *The Moahunter Period of Maori Culture*, 2nd ed. (Wellington, 1956) pp. 73, 271–2; also in J. Golson, 'New Zealand archaeology, 1957', *Journ. Polynes. Soc.*, LXVI 3 (Sep 1957) 287. See also D. Teviotdale, 'Excavation of a moahunters' camp near the mouth of the Waitaki River', *Journ. Polynes. Soc.*, XLVIII 1 (Mar 1939) 161–72; and J. Lindsay Buick, *The Moahunters of New Zealand* (New Plymouth, 1937) p. 189.

the moahunter would find that his fire would carry, especially on the drier slopes of the Central Otago ranges, on the foothills and over the loess-mantled downlands. Elsewhere his success would be limited to forest clearance by patch burning, along the streams and river flats and in the damper valleys and plains of Southland, on the exposed and windy spurs of Banks Peninsula and on the deep and damp sub-coastal soils of the Canterbury Plains. In such localities, some forest withstood his depredations; and in some damper situations he only succeeded in enabling one type of tree growth to replace another, in creating culturally-induced vegetational 'discontinuities', in disturbing the environment of the remnant matai stands so that they gradually decayed, in reducing forest types to disturbed 'pocket hand-kerchief' stands, in producing 'a most intricate and kaleidoscopic confusion of various forest types', and in becoming responsible for a vegetation, east of the alpine ranges, which Holloway describes as 'a gross discontinuity' underlying which 'can be traced no uniform factor of climate or of soil' (p. 355).

FIRE, FOREST AND GRASSLAND

Carl Sauer has long insisted and recently re-emphasised[1] that the climatic origin of grasslands 'rests on a poorly founded hypothesis'. Holloway himself, the most ardent advocate of the thesis relating the occurrence of grassland in the South Island to climatic change, provides evidence to support Sauer's contention in respect of New Zealand's tussocklands. As Holloway admits (p. 355), both the grasslands and the residual patches of matai forest extend over a wide climatic gradient at least from 80 in. of annual average precipitation to less than 15 in.; from winter-quarter to summer-quarter maxima of rainfall; and from Thornthwaite's AC'r to DC'd climates).[2]

Throughout the pre-European tussock grassland, and more especially in gullies, on terrace edges and wherever the surface configuration of the plains was broken, woody plants occurred, including occasionally surviving residues of podocarp forest understoreys.[3] And where fire is prevented and initial protection from stock is

---

[1] Sauer, in Thomas, p. 55.

[2] B. J. Garnier, 'The climates of New Zealand according to Thornthwaite's classification', *Ann. Assoc. Amer. Geogrs.*, xxxvi 3 (Sep 1946) 141–77.

[3] V. D. Zotov, 'Survey of the tussock-grasslands of the South Island, New Zealand', *N.Z. Journ. Sci. & Tech.*, xx 4A (1938) 212–44.

provided, the plains and the drier, frostier inland basins support large exotic trees. Indeed the tussocklands are criss-crossed with maturing shelterbelts and accommodate exotic forest plantations. Even on elevated ridges in Central Otago, facing the sun and exposed to the nor'wester, and in North Canterbury, totara has not only survived all the fires and grazing and all the alleged deterioration of climate, but is regenerating vigorously and invading 'depleted tussock steppe'. Nor is the extent of tussock vegetation determined by pedologic or lithologic factors. On hard rock highland, schist mountain slope, gravel plain, loess-spread downland, limestone hill and basalt ridge, the tussock associations are equally at home. Position and extent reveal a rather closer relationship to surface configuration, forest sometimes taking over at the break of slope. This is particularly true of beech 'hangers' and tussock 'parkees' in the high country, as well as along the inner rim of the Canterbury Plains. It may be no coincidence that with contemporary fires, damage also often stops short at barriers presented by breaks of terrain.

EVIDENCE OF ANCIENT FIRES

So far this association of moa, moahunter and incendiarism has been largely speculative and imaginary. Paradoxically the best collection of supporting evidence to lend substance to the argument is to be found in Holloway and Johnston. It is reinforced by other fieldworkers and by Maori traditions.

Holloway states (p. 373) 'that considerable fires did rage is indisputable', and (p. 362) that moahunters 'used fires [and] had no effective means of controlling fire, and existing forests carry the mark of these ancient fires to this day . . . man was an agent in forest destruction'. But – and it is a big 'but' – 'climatic change occurred either before or after the fires. *It could only have been before*' (p. 373; present writer's italics). 'The destruction of the forests by fire apparently followed closely on the heels of a climatic change' (p. 374).

Whilst he agrees that the final destruction of the podocarp forests was accomplished by fire, he does so almost by way of incidental asides. Holloway's case is that a postulated recent change of climate alone can bring to order the 'patchwork confusion' in which he finds the patterns of South Island forest vegetation. Consequently when, in any particular forest situation, the theory fails to explain the anomalous facts, Holloway's resort is to some 'peculiarly favourable local site factor complex' or to soil, surface form, geology, lithology or unnamed 'edaphic factors'.[1] In no individual case where anomalies are to be explained does he allow fire or cultural interference (except only since European settlement) to be ranked as a 'factor' in the explanation of why the vegetation is not attuned to the present climate. (Compare pp. 337–8 and p. 346 where possible 'factors' are reviewed.) Although 'old matai/kahikatea and Hall's totara/ kaikawaka forests were [ultimately] destroyed by fire', and although 'the final decay and breaking up of the forest was undoubtedly accelerated by fire' (p. 361), Holloway fails to admit to the list of factors controlling and conditioning the processes of forest evolution fires caused by early man whether by accident or by design. Nor does he consider the possibility (in the South Island) of fires kindled by natural agency such as lightning.[2]

He does, however, make incidental reference to fallen totara logs 'frequently charred on the outside' (p. 373), to 'beds of charcoals' on top of 'truncated [?] fossil soils strongly leached' where, at 3000 ft in the Takitimu Mountains, 'forest was destroyed by fire' (p. 263), to 'charcoals in subsoil', and to 'a fire history [on the 'highly inflammable' Grey Valley *pakihis*] dating back to pre-European days' (p. 391). Together with his frequent amazement at the 'abruptness' of forest-type boundaries (p. 336), his admission of the beech's ability to colonise *bare* ground whilst finding it difficult to penetrate standing podocarp forest (p. 338), his reference to beeches occupying what appear to be former 'clearings' in the matai stands (p. 347), his detailing of the ultimate encroachment of well-known pyrophytes

[1] Even in the Grove Burn, Alton Burn and Lill Burn valleys of western Southland, where the theory of climatic change was developed, this is the case. For example (p. 340), 'The ridge crest podocarp stands and the podocarp forest on the warm Grove Burn slope are still moderately well in harmony with local climates [although on cold slopes and in valley bottoms they are in decay, or have been replaced]. The *aspect* is exceptionally favourable in the one case while, in the other, the more favourable soil conditions found on the ridge crest gravels may tend to outweigh climatic factors.'

[2] Indeed Holloway is contradictory, for, whilst admitting to widespread destruction by fire, if only after climatic change (pp. 372–4), he later (p. 402) claims that the Polynesians possessed 'but a stone age culture' and 'were typically food gatherers and hunters and, as such, *left little mark on the forests* as a whole'. This he argues to counter the quoted reference to the fact that in Europe the effects of climatic change are 'entirely masked or overriden by changes consequential on the activities of men or of animals'. Holloway underrates the role of fire set by New Zealand's first culture group and ignores the effects of grazing by moas, or of fires lit by nature.

(tussock grasses, cabbage trees and manuka) (pp. 359, 360 and 361): all add up to formidable evidence supporting the claim that fires occurred widely and frequently and must inevitably have had a role of some importance in determining the pattern and character of the plant cover.

The traditional Maori references to the 'Fires of Tamatea' have been frequently reviewed and frequently interpreted.[1] In the belief that the Polynesian was always, like the classic post-Fleet Maori, a conservationist and a protector of Tane, most students of the Maori interpreted these mythical conflagrations as referring to either a sacred, religious flame, or to the fires occasioned in dim and distant times by volcanic activity in the North Island, despite the fact that in Maori story and legend Tamatea's fire is nearly always associated with burning the vegetation cover and with the complete destruction or near-elimination of the moa.

The memory of the fires of Tamatea are as indistinct in the recollection of post-Fleet tradition as are most genuine accounts of the moa. Both, in any case, are clearer than any traditional account of a change of weather and climate. Tamatea, the incendiarist, was thought to have arrived with the Fleet; but Downes has demonstrated that there were at least four earlier Tamateas in the twelfth and thirteenth centuries, descended from Toi, and that 'not a word about the fire myth' is associated with the particular history of the Fleet navigator.[2] Moreover, other legends suggest that even 'in the days of Kupe [A.D. 950] the first folk who came hither lit fires at all the places they landed at'.[3] It seems clear that the fire of Tamatea could well be the fires that destroyed the matai forests which Johnston's collection of C[14] dates[4] indicates to have taken place largely between A.D. 1075 and 1290. These are the extreme dates he cites of the logs and burning of which was, according to C[14] analysis, '*coincident with death*'. Firing of grass and scrub, and indeed of more accessible forest, must

---

[1] See, for example, Duff, Holloway, Buick, Lockerbie, etc., in references cited above.

[2] T. W. Downes, 'New light on the period of the extinction of the moa', *Trans. N.Z. Inst.*, XLVIII (2) (1916) 426, 431–3.

[3] Elsdon Best, 'The forest lore of the Maori', *Dom. Mus. Bull.*, no. 14 (1942) 235. Duff in *The Moa-hunter Period of Maori Culture* reviews in detail the traditional references to the moa and, incidentally, to the 'Fires of Tamatea' with which their 'loss', 'destruction', or 'extinction' is causally associated.

[4] J. A. Johnston, 'Recent Climatic Change in South Island, New Zealand: A Geographic Analysis', University of New Zealand M.A. Thesis (University of Canterbury), unpublished, 1958.

have taken place even before A.D. 1075 and no doubt was continued in the fourteenth and fifteenth centuries as the moas became very scarce.

The fires of Tamatea did not finally exterminate the moa; but they did demonstrably destroy the forest and extend the area of grassland. This they did earlier than has hitherto been suspected. They raged in Central Otago and North Canterbury in the widely postulated pluvial period in the eleventh and twelfth centuries and considerably before the date of desiccation propounded by the strongest advocates of climatic change.[1] The moa survived these fires, but probably in reduced numbers. According to Rangitane tradition the *kuranui* was 'lost or destroyed',[2] which accords with other evidence that the diminishing population of flightless birds – especially of the larger species – retreated southwards, and into the bush where in regions with heavier precipitation they were subsequently harried and hunted by further firing – probably patch-burning – during the next three or four centuries.

Destruction of forest by cultural interference, rather than by insidious climatic change, would bring about a sudden and much more effective transformation of the vegetative cover. This would be a more effective 'trigger' than the slow operation of a climatic fluctuation. If, as is suggested here, the fires were first lit towards the end of a relatively warm and dry oscillation of weather sequences (rather than before the onset of desiccation postulated by a succession of recent commentators), the damage to a virgin plant cover could well have been precipitous and widespread. As was the case with forest destruction in the much damper and more humid rain forest districts of the North Island centuries later, and without the benefit of surface sowing of an exotic replacement cover, accelerated erosion would follow. Only with such immediate and calamitous consequences is it conceivable that standing forest near Christchurch could have been buried in shingle by the Waimakariri as early as A.D. 1190. If the forest destruction which led to this deposition of shingle was occasioned by climatic change, that change must be dated earlier than has hitherto been suggested, must have occurred

---

[1] But compare Holloway ('Forests and Climates in the South Island . . .', p. 373) '. . . without question the forests were not burnt until after they had entered a period of instability as a consequence of climatic change . . . [it] could only have been before [the fires].'

[2] Quoted by Johnston, 'Recent Climatic Changes in the South Island', p. 38.

in watershed regions where the climate was not particularly 'sensitive' and must have been abrupt, sudden and 'cataclysmic'. For, before erosion could have been caused on the scale contemplated, not only must regeneration have ceased but the whole forest association, including undergrowth and mature trees, must have been killed within a few years at most by 'desiccation', or cold, leaving a wilderness of dead trunks and bare eroding ground between. Climatic change, however, does not operate like this.

Cultural interference and the widespread and repeated use of the firestick is an adequate explanation of the facts – and even many of the anomalies – revealed by the adherents of the hypothesis of climatic change. It is not necessary to invoke a change of climate, whether unnaturally abrupt or normally insidious, or whether it is dated A.D. 1200, 1300, or after the fourteenth century.

FOREST SURVIVAL AND REGENERATION

After careful and detailed consideration of climate as well as close field investigation, Zotov claimed over twenty years ago that much of the tussock grassland was 'induced'. 'Without interference it would gradually pass into scrub and finally back into forest which is the climax formation naturally occurring under existing climatic conditions.'[1] Johnston's evidence from the Pisa Range illustrates the validity of this claim even when applied to elevations up to 3000 ft and to very dry and exposed slopes. Zotov also delimited on a climatic basis the limits of 'true' as opposed to 'induced steppe'. This is reproduced in Figs. 17.1 and 17.2 in which an attempt has also been made to map the extent of the podocarp forests before the fires of Tamatea. Zotov points also to the woody forest species that occurred in the grassland on the advent of the European pastoralist, and to those that frequently appear even today.

The eleventh- and twelfth-century conflagrations would be followed by further fires. Conditions were made highly suitable for pyrophytes, amongst which grasses were the first colonisers. Even if forest species and sources were still available, the intravegetational and soil climates would be so completely transformed by fire that in many areas, where the contemporary atmospheric climate was borderline for the forest associations, regeneration would be slow

[1] Zotov, 'Survey of the Tussock Grasslands . . .', p. 213.

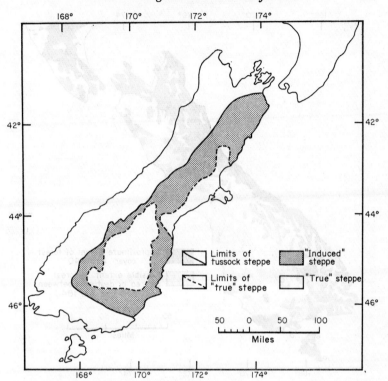

Fig. 17.1   Distribution and limits of 'true' and 'induced' steppe according to Zotov

and difficult.[1] Repeated fires made it impossible; although, as Holloway shows, the beeches revealed in some districts a capacity to colonise bare ground and mineral soils.

But in damper districts and in the later absence of fire there is evidence of regeneration of forest in grassland areas. It does not rest on Zotov, or Johnston alone. The early testimony of Buchanan as to the nature of the vegetation early in the period of European colonisation is important. 'On many grassy ridges [in Central Otago] may be found the remains of large trees, and over large areas the surface is dotted with the little hillocks and corresponding hollows produced by the upturned roots of trees which had been blown over

[1] There are sharp differences between intravegetational and soil climates before and after burning and after other forms of cultural interference with the vegetation.

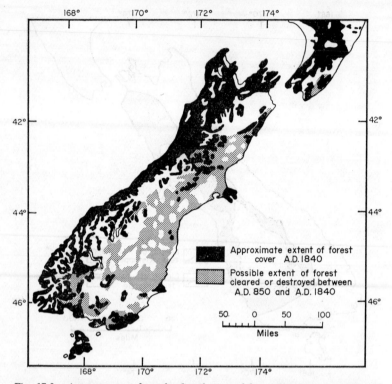

Fig. 17.2 An attempt to show the distribution of forest vegetation in 1840 and the extent of the forest vegetation destroyed (probably largely by burning) in the preceding millennium. Based on N.Z. Forest Service, A Descriptive Atlas of New Zealand, Holloway and Johnston.

generally in the line of prevailing winds, after their *destruction by fire.* . . . At the beginning of [European] settlement large tracts of the Province [of Otago] *were being reclothed with bush*, but as the country was opened for cattle and sheepruns this new growth was *again* burned off.[1]

[1] Buchanan, 'Sketch of the Botany of Otago', p. 181 (italics the present author's).

18 The Eastern Grassveld Region

J. W. BEWS

From *The Grasses and Grasslands of South Africa* (P. Davis & Sons Ltd, Pietermaritzburg, 1918) pp. 116–17, 127–9.

STABLE GRASSLAND – CLIMAX STAGES

IN the later stages of development a large number of species take part, the majority belonging to the tribe Andropogoneae, which includes *Anthistiria imberbis* (inSinde or Red Grass), the species which is usually dominant. When we compare these subsequent species with the pioneers, we notice many points of difference. The former are not so xerophytic, nor so deep-rooted. They form a closer mat over the surface of the soil, and thus prevent the run-off of water to a large extent. Their shallow spreading root systems obtain moisture first, before it is able to soak down to the lower levels, and this is partly why they are able to oust the pioneers. Light, however, is another factor of considerable importance. The seedlings of *Anthistiria*, for instance, prefer to germinate in the shade of other herbage. It is particularly interesting to notice in mixed transitional grassland how young plants of *Anthistiria* almost invariably come up through the middle of a tuft of *Aristida*, which gradually becomes smothered as the *Anthistiria* grows taller and shades it, and also forms a mass of roots immediately above those of *Aristida*. *Aristida* seedlings themselves, on the other hand, do not seem to tolerate shade. They usually germinate in the open spaces. The pioneer species therefore prepare the way – and it is a necessary preparation – for the establishment of the climax types.

Over by far the largest areas of Eastern grassland *Anthistiria imberbis* (inSinde) is the dominant species. It covers enormous areas in the Eastern Province of the Cape, Natal, the Northern and Eastern Free State, and the Transvaal. There are several varieties, which are not always easy to distinguish. A closer study of the species would repay attention, and ought to be undertaken. Stapf has distinguished three varieties, viz., *mollicoma* with leaves and involucral spikelets densely hairy, *argentea* with leaves densely hairy, but involucral spikelets glabrous or scantily hairy, and *burchellii* with laxer panicles and longer spathes, but there are certainly many intermediate varieties, and the different extremes may be found growing together. Though I have given a good deal of attention to the species, at present I prefer not to record any observations on the distribution of the separate varieties. *Anthistiria* is the chief pasture grass in South Africa, and it makes a good hay. Its innovation buds are intravaginal, and it is consequently easily burnt out. Its dead leaves tend to break up into fibres which form a spongy mat over the surface of the soil, thereby retaining water, and lessening soil erosion. As already mentioned, it is not so deep-rooted as the species which precede it.

THE TRANSITION TO WOODLAND
(POST-CLIMAX GRASSLAND)

A great deal of the Eastern grassland at the present time occupies what are natural forest climatic habitats. The south-eastern slopes of the High Veld in Natal very commonly do bear forests if they have not been destroyed, and over most of the High Veld timber trees may be successfully grown. If Grassveld, which occupies such forest areas, is left unburnt for a number of years, the early stages of a forest sere soon make an appearance.

The first stage is usually the replacing of *Andropogon-Anthistiria* Veld by taller *Andropogon* species of the Cymbopogon section – the grasses which are commonly referred to as Tambookie. The Tambookie consocies that is most frequent is one dominated by *A. nardus* var. *validus*, a lemon-scented grass known to the natives of Natal as isiQunga and used by them medicinally. *A. dregeanus* (uQunga), another equally tall species, is also common, and *A. auctus* (also called uQunga), which is rather like *A. hirtus*, but taller, is a third member of the associes. The three species are often mixed. At the present time attention is being directed to them in connection with the possible establishing of a paper-making industry in the Union. Other species of *Andropogon* which belong to the same transitional stage of the succession are not so common, e.g. *A. plurinodis*, *A. rufus*, *A. dichroos* (a spring-flowering species), *A. filipendulus*, *A. cymbarius* and its variety *lepidus*. The Tambookie and other taller Cymbopogon associes are not confined to the early stages of the forest sere. They are also characteristic of the hydrosere and often occur in the sub-sere, where land has been cultivated.

There are several other grasses which are characteristic of the early forest stages: *Erianthus capensis* (umTala) and *E. sorghum*, *Arundinella ecklonii*, *Phalaris arundinacea*, *Setaria sulcata*, *S. lindenbergiana*, *S. nigrirostis*, *S. flabellata*, *S. aurea*, *Pennisetum unisetum*, and occasionally other of the Vlei Pennisetums, *Panicum crus-pavonis* and sometimes other Panicums *Sporobolus rehmanni*, most of which also belong to the hydrosere.

In the transition to forest the grasses soon begin to give way to shrubs or other flowering plants. The Bracken fern, too (*Pteris aquilina*), sometimes covers large areas to the exclusion of everything else. Composites like *Athanasia acerosa*, *Artemesia afra*, *Berkheya platyptera* often form quite large consocies. At the higher altitudes, *Myrsine africana* is a common forest pioneer. *Leucosidea sericea* gradually establishes definite *Leucosidea* or Oudehout Scrub, which in turn progresses towards Yellow-wood (*Podocarpus*) forest. *Buddleia salviaefolia* is a very frequent precursor of scrub and forest also, all over the midlands and mountain regions of Natal. At Nottingham Road, Natal, on the farm of Mr James King, there is an area of over 60 acres which has been protected from grass fires by fire-breaks for over thirty-five years. Through the kindness of Mr. King, I have been able to make a careful study of it. The various stages of the succession are clearly shown. First, Tambookie grasses, the species named above. Second, *Athanasia*. Third, *Buddleia salviaefolia*, *Leucosidea sericea*, *Erica cooperi*, species of *Rhus* and *Lasiosiphon*. Fourth, young Yellow-wood Bush. There is no *Anthistiria* Veld left, though I am assured by Mr King that the inSinde used to be dominant over the whole area.

Farmers whose farms lie in the forest climatic areas will usually find that, if they refrain from burning the grasses of their Veld, the succession progresses in the same direction. The grasses grow taller and coarser, and gradually the finer grasses are ousted. Burning then becomes necessary to keep back the succession. It may be urged that such farmers would probably find the planting of timber a better financial proposition than the pasturing of cattle, but if they must have pasturage, then they must continue to burn the grasses. Forest soils are usually not rich, and the grasses do not as a rule give a good pasturage at any stage of the succession.

There are other transitions from grassland to scrub, which follow different lines. In the Low Veld, and in all the various types of Bush Veld, etc., the succession is that given in detail in one of my former papers. Isolated trees, usually acacia thorn trees, germinate among the

Anthistiria–Andropogon grasses. They grow up and form a park-like type of vegetation. Birds alight on their branches, and bring the seeds of numerous other species, which are deposited underneath the pioneer trees, and grow up in their shade. Clumps of trees are thus formed, and as the clumps grow closer thorn thickets are established. The climax type is scrub. For full details, see *S.A. Journal of Science* (Nov 1917), where a list of over 230 species of Thorn Veld trees and shrubs is given, with an account of their relative frequency and behaviour in the plant succession.

19 The Sonoran Desert

J. R. HASTINGS and R. M. TURNER

From *The Changing Mile* (University of Arizona Press, Tucson, 1965) pp. 185–9.

JUDGED either by the diversity of life-forms to be found in it, or by the number of species making up its flora, the Sonoran Desert supports the most complex vegetation of the four arid regions of North America. At its most varied, in the upland situations of southern Arizona where broken terrain produces a variety of microclimates and where coarse, rocky soils in conjunction with well-developed drainage patterns give rise to similarly heterogeneous edaphic environments, a single vicinity may contain plants representing more than half of the twenty-five life-forms suggested by Shreve [1] as a basis for classifying desert plant life. Throughout the mosaic the prevailing stress may be that of moisture deficiency. But the ways are many by which species have adapted to this pre-eminent condition through modification to anatomy, physiology or morphology. And although a single community, occupying a single patch of the mosaic, may include only a few species, there are so many communities that the total number of plants growing in the Sonoran Desert is large indeed.

On the basis of characteristics shared by the dominant members of the vegetation, Shreve [1] has subdivided the desert proper into seven regions that can be conveniently referred to as provinces. In general, broad climatic and topographic factors shape the provincial boundaries, but within the bounds of a single province the vegetation may range from simple to complex depending on variations in soil and microclimate. Descending the gentle outwash slope, or bajada, which surrounds a desert mountain range, one finds a progressive change towards simplicity as the terrain becomes smoother, as the soil tends to become finer and more uniform, and as the gradual loss of a well-defined drainage pattern all tend to produce homogeneous soil conditions. Between the upper part of the bajada and the lower, many perennial species drop out; only a few new ones enter. The life-forms become fewer in number, with a marked tendency in the lowlands for those evergreen plants to become dominant that are capable of biseasonal growth.

Some of the effects of microclimate have already been discussed.

The importance of soil in determining the character of desert vegetation resides in the strong control it exercises over moisture through its regulation both of the quantity of water available and the duration over which moisture is present [1]. Although much of the work in the North American deserts has stressed the importance of salt content, Shreve himself made little of this soil factor, stressing instead such physical features as texture, depth, and surface characteristics.

Yang and Lowe [2] have shown that the relatively level sites at the bases of bajadas are more xeric than the slopes themselves because of soil texture differences.

The effect of soil depth may be seen in the reduction of perennial plant cover to values of 5 per cent or less where the relatively coarse, absorbent, upper layers of soil are shallowly underlain by impervious horizons of caliche or of hardpan. Here, the downward movement of soil moisture is impeded, and much of the water that might otherwise be available to plants is lost by evaporation.

Surface configuration may influence plant life either by inducing heterogeneous vegetation where the soil surface is dissected and irregular, or by inducing relatively simple plant cover where level terrain, undissected by runnels, imposes more uniform soil-moisture conditions. Perhaps nowhere in the Sonoran Desert are differences so stark between the plant life of these two kinds of terrain as they are in the first of the major provinces, the Arizona Uplands.

THE ARIZONA UPLANDS

By reason of the relative abundance of moisture they receive and the wide range of elevations they span, the Arizona Uplands are the most diverse of the three provinces of the Sonoran Desert dealt with. They lie along the north-eastern edge of the desert region (Fig. 19.1), where true desert vegetation may extend upward to elevations above 3000 ft on warm, south-facing slopes. From that approximate upper limit, the plant life of the province extends downward towards the west and south to elevations of about 1000 ft. Average annual rainfall varies from 7 in. to 12 in., the amount being closely dependent on elevation [3]. Mean annual temperatures range roughly from 64°F. to 72°F.

A bajada may once again be used to illustrate the variety of vegetation to be found. On the higher reaches of the slope 40 per cent or more of the surface may be covered by the crowns of woody and

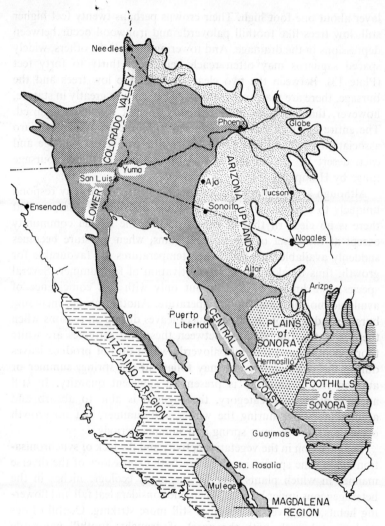

Fig. 19.1 *The vegetation provinces of the Sonoran Desert*
(*after Shreve* [1])

succulent perennials. Here desert growth reaches its most luxuriant stage, often producing a cover through which it is difficult to see for more than a few hundred feet. Beneath the low shrubs, sometimes occurring with marked fidelity, one finds such small plants as pincushion cactus. Above these, bursage commonly grows, in a uniform layer about one foot high. Their crowns perhaps twenty feet higher still, low trees like foothill paloverde and ironwood occur between depressions in the drainage. And towering over all the others, widely spaced saguaros may often reach heights of thirty to forty feet (Plate 13). Between the two clear strata of the low trees and the bursage, there are many other plants. They vary so greatly in stature, however, that no layer of intermediate height can be recognised. The entire complex has been referred to as the paloverde–saguaro association by Yang and Lowe [2], as the paloverde, bursage and cacti desert by Nichol [4], and as the paloverde–triangle bursage range by Humphrey [5].

Although several plant species occur together, each may respond uniquely to seasonal changes in moisture and temperature, so that there is no clear rhythm of activity within the bajada community except at the onset of the summer rains, when moisture becomes suddenly available at a time when temperatures are favourable for growth, thus breaking the long aestivation of foresummer. Several species, like limber bush, leaf out only with this coincidence of available moisture and high temperature. Another group, including bursage and creosote bush, produces leaves during all seasons when there is water. Intermediate between these two extremes are white thorn, ocotillo and foothill paloverde which do not produce leaves during the cold winter, but may foliate in the spring, summer or autumn if soil moisture is present in sufficient quantity. In still another phenologic category, the saguaro is able to absorb and accumulate water during the winter and summer, but no growth occurs until the warmer spring and summer periods.

The variation in the vegetative activity and the lack of synchronisation among the species involved are both expressions of the diverse manner in which plants may fill the many ecologic niches in the heterogeneous upper bajada. When one considers leaf fall and flowering habits, the diversity becomes still more striking. Ocotillo loses its leaves abruptly with the onset of drought; foothill paloverde retains its foliage over longer periods of dryness. Bursage loses its leaves slowly, the older ones first and the younger ones later, until with prolonged drought only the very youngest leaves, at the stem tips, remain; these too may die if desiccation is extreme.

Flowering is often less dependent upon rainfall than is vegetative growth. Particularly among the cacti, flowers may appear without regard to moisture conditions; thus, the saguaro blooms unfailingly each year, the flowers usually opening during May in advance of the summer rains. Foothill paloverde does not bloom each year, and its failure to do so is apparently related to insufficient winter rainfall; yet, in years when floral production does occur, flowering comes in May during the arid foresummer [1, 6]. Other plants, such as creosote bush, may produce flowers during any month of the year when moisture is adequate.

A second conspicuous community of the bajadas occurs in narrow, branching ribbons along the drainage channels that interrupt the paloverde–saguaro community, and is analogous to the gallery forests of higher elevations. White thorn and catclaw are most typical here, their size depending upon the amount of moisture available. Where minor washes coalesce in a single large channel, these plants reach heights of 15 to 20 ft and may be joined by mesquite, blue paloverde, desert willow, canyon ragweed, and other species requiring the improved moisture balance of this habitat. Several of these desert riparian species are among those plants known to have invaded non-riparian positions above the Sonoran Desert in the desert grassland and oak woodland.

As the base of a bajada is approached, many perennials of the upper slope are no longer present; ironwood, foothill paloverde bursage and saguaro, among others, are gradually lost in the descent. Towards their lower limit, the individuals of these species become confined to small drainage ways, with the intervening areas now occupied by creosote bush and white bursage, members of the community from the plain below.

This third community is low in stature and simple in composition, two features which bespeak the less favourable moisture supply to be found on the basal plain. Creosote bush is the principal dominant, occurring as a widely and rather uniformly spaced plant about two feet high in the drier habitats and over six feet high where moisture is more abundant. Under optimum conditions it may attain coverage values of 15 to 20 per cent. White bursage, its principal associate, occupies the broad openings among the larger shrubs. It rarely exceeds two feet in height and may have coverage values of from less than 1 per cent to 10 per cent. Total coverage for the community may vary from 15 to 30 per cent, depending on the availability of soil moisture. Unlike the paloverde–saguaro association, the creosote bush – white bursage community is not restricted to the Sonoran Desert, but occurs along the valleys of the Mohave Desert as well [1, 7].

A similar distribution characterises still a fourth plant community of wide extent in the Uplands. Although restricted in the Sonoran Desert to the northern part, areas dominated by desert saltbush, like those dominated by creosote bush and white bursage, also extend into the Mohave Desert. It occurs in essentially pure stands on a variety of soils: on well-drained, fine, sandy loam; on other types where surface drainage is impeded [8]. The total plant coverage expresses the relative water balance of each location; coverage values as low as 3·5 per cent have been noted in a site where downward percolation is blocked by haidpan. As with many plants of the desert plains, vegetative growth occurs during both rainy seasons; flowering is restricted to autumn. According to Aldous and Shantz [9], the dominant vegetation of the bottom lands along the Gila river is also desert saltbush. Thus the association extends from the Arizona Uplands through the second of the provinces, the Lower Colorado Valley.

THE LOWER COLORADO VALLEY

Proceeding westwards along a line connecting Tucson, Arizona, with San Luis, on the Colorado river in Mexico, one crosses a series of plains that descend to the river like giant stairs, each separated from the other by mountain ridges that grow lower and less massive the farther one travels from the backbone of the continent. With each tread the plant life grows more impoverished; the simple creosote bush – white bursage community is no longer confined to the valleys; where foothill paloverde, bursage, ironwood and saguaro do occur, they, in contrast, are restricted almost entirely to drainage ways. Beginning about at Ajo the vegetation of the Arizona Uplands grades into that of the Lower Colorado Valley.

The final 250 miles of the Colorado river flow through this low, arid province, which extends eastward along the course of the lower Gila for another 200 miles. To the west it reaches 150 miles into the Imperial Valley of California; to the south, narrow extensions into Mexico flank the head of the Gulf of California.

Throughout its extent, the province is confined mainly to elevations lower than 2000 ft, and in the Imperial Valley it extends down to well below sea-level. It is at once the hottest and the driest of the desert subdivisions. Mean annual temperatures for stations within it range from 65°F to 74°F; the average precipitation from slightly more than 1 in. per year to slightly less than 8 [10, 11, 12]. Biseasonal in distribution, the rainfall tends to be equally distributed between winter and summer or, in the westernmost reaches, slightly unbalanced in favour of winter.

As one might expect from the degree of aridity, the vegetation is simple, sparse, and relatively uniform. The commonest community is one dominated by creosote bush and white bursage. Here, in contrast to their distribution in the Arizona Uplands, these two species are not confined to valleys. They cover vast stretches of plains, bajadas, and even volcanic hills, their density and height varying with the amount of moisture available from the local soil. On some volcanic outcrops a depauperate phase of the paloverde–saguaro community of the Arizona Uplands may be present instead. Under these conditions, foothill paloverdes characteristically are low and widely spaced; and as one approaches the Colorado river, the infrequent saguaros disappear almost completely.

The runnel vegetation of the province includes foothill paloverde and ironwood, but lacks white thorn, one of the more conspicuous plants in similar habitats in the Uplands. Along some of the broader washes mesquite and blue paloverde occur. The smoketree, absent from most of the Arizona Uplands because of its sensitivity to cold, grows here in riparian situations [13]; conversely desert willow is absent below about 1500 ft [13], although it is abundant along washes in the province to the east, and in the grasslands.

Expanses of stabilised sand characterise large areas of the Lower Colorado Valley, and the typical community on these is dominated by big galleta, growing either in pure stands or in company with creosote bush and other perennials. The apparent anomaly of a grass-dominated community existing under a climate favourable to shrubs and low trees may be best understood through a knowledge of this plant's life-form. Unlike most grasses, big galleta has perennating buds borne above the soil surface on woody culms, a characteristic shared with many desert shrubs. Although its phyletic relationship is with the grasses, areas in which it is the dominant plant are not analogous to grasslands.

In its physiography the Lower Colorado Valley is more uniform than the other two provinces considered; only a few mountain ranges, and they relatively low, interrupt the plains. Sand dunes and malpais fields – the latter the product of recent volcanic activity – occur towards the south and support distinctive communities of their own. The Pinacate Mountain region, where lava flow and dune exist side by side, is a chequerboard of dark and light localities that support an intricate mosaic of vegetation responding sharply to the shifting patterns of soil and albedo. Pockets of relatively deep sand support big galleta; other sandy areas are dominated by ocotillo, and in some cases even by saguaro; the volcanic hills maintain a sparse cover of foothill paloverde, ironwood, elephant tree and *Jatropha cuneata*. The latter two plants have affinities to the south where, with their fleshy-stemmed relatives, they form the characteristic vegetation of the third and last of the provinces dealt with, the Central Gulf Coast.

THE CENTRAL GULF COAST

The vegetation of the Central Gulf Coast occurs in two coastal strips along opposite sides of the Gulf of California; to the north both give way to the Lower Colorado Valley and today, if not in the past, the two strips make no contact with each other. The province has been described by Shreve [1] as the driest in the Sonoran Desert, but from the weather records available now this distinction clearly belongs to the Lower Colorado Valley [11, 12]. The coastal region seems to be characterised more than anything else by rapidly changing gradients of rainfall. The isohyets are closely crowded, more or less paralleling the coastline, and the amounts fall precipitously as one approaches the Gulf from the east. Guaymas, one of the few stations with a long-term record, receives an annual average of about 9·5 in., with 75 per cent falling during the six hot months [12]. This amount is probably near maximum for the province; most parts receive between 4 and 6 in.

In contrast to the two provinces already described, the vegetation of the Central Gulf Coast lacks the sharp distinction between communities of the plains on the one hand and the bajadas and hills on the other. As the traveller approaches the region from the Lower Colorado Valley, creosote bush and white bursage become sporadic in their occurrence, and no longer dominate extensive areas. Hills and intermontane plains, valleys, and bajadas alike are covered by open communities of arborescent or shrubby physiognomy.

Relative to the Arizona Uplands, foothill paloverde loses its dominant position in the vegetation, and ironwood and blue paloverde gain in importance. Elephant tree and *Jatropha cuneata*, plants of limited occurrence in the Lower Colorado Valley, are joined by a host of other fleshy-stemmed, sometimes aromatic plants: *Bursera hindsiana, Jatropha cinerea, Euphorbia misera*. In most situations, none of the perennials asserts dominance to the degree that foothill paloverde, for example, does at places in the Arizona Uplands. The saguaro is of minor importance, its place in the vegetation being occupied now by the *cardón*, or by other columnar cacti like the organpipe cactus and *sinita*.

The boojum tree, another fleshy-stemmed species, is found at only one location on the Mexican mainland, but occurs abundantly there. Its maximum age, judging by its growth in the interval between pictures taken in 1931–2 and 1963, must be close to 400 years; thus it ranks among the longest-lived of the Sonoran Desert plants for which such information is available. The mainland occurrence of this grotesque tree has been described by Shreve [1] as covering only a few square miles in the vicinity of Puerto Libertad (Punto Kino or Punto Cirio); Aschmann [14], however, extends its range nearly as far south as Desemboque, and he is evidently correct. The disjunct distribution of the boojum tree from its present centre of occurrence in Baja California to the single, minor outpost on the mainland remains one of the many intriguing problems involving the flora of the Sonoran Desert.

An important community, found also in other provinces of the desert with coastal contacts, is dominated by *Frankenia palmeri*. Often associated with *Atriplex barclayana*, it dominates a low, wind-swept community that typically is found just inland from the strand. On sandy soils the monotonous appearance of the association may be broken by taller plants found abundantly inland as well: *Jatropha cuneata, J. cinerea, Euphorbia misera*, jojoba, and teddybear cholla.

REFERENCES

[1] SHREVE, FORREST (1964) 'Vegetation of the Sonoran Desert', in Forrest Shreve and Ira L. Wiggins, *Vegetation and Flora of the Sonoran Desert*, 2 vols (Stanford, Calif.: Stanford University Press).

[2] YANG, T. W., and LOWE, C. H., JR (1956) 'Correlation of major vegetation climaxes with soil characteristics in the Sonoran Desert', *Science*, CXXIII 542.

[3] GREEN, CHRISTINE R. (1959). 'Arizona statewide rainfall', *Technical Reports on the Meteorology and Climatology of Arid Regions*, no. 7 (University of Arizona, Institute of Atmospheric Physics).

[4] NICHOL, A. A. (1952) 'The natural vegetation of Arizona', *University of Arizona Agricultural Experiment Station Technical Bulletin*, 127.

[5] HUMPHREY, R. R. (1960). 'Forage production on Arizona ranges, V: Pima, Pinal and Santa Cruz Counties', *University of Arizona Agricultural Experiment Station Bulletin*, 302.

[6] TURNER, R. M. (1963) 'Growth in four species of Sonoran Desert trees', *Ecology*, XLIV 760–5.

[7] ALLRED, D. M., BECK, D. E., and JORGENSEN, C. D. (1963). 'Biotic communities of the Nevada test site', *Brigham Young University Science Bulletin*, Biological Series II, no. 2.

[8] SHANTZ, H. L., and PIEMEISEL, R. L. (1924) 'Indicator significance of the natural vegetation of the South-western desert region', *Journal of Agricultural Research*, XXVIII, 721–801.

[9] ALDOUS, A. E., and SHANTZ, H. L. (1924) 'Types of vegetation in the semiarid portion of the United States and their economic significance', *Journal of Agricultural Research*, XXVIII 99–127.

[10] GREEN, CHRISTINE R., and SELLERS, W. D. (eds) (1964) *Arizona Climate* (Tucson: University of Arizona Press).

[11] HASTINGS, J. R. (1964a) 'Climatological data for Baja California', *Technical Reports on the Meteorology and Climatology of Arid Regions*, no. 14 (University of Arizona, Institute of Atmospheric Physics).

[12] —— (1964b) 'Climatological data for Sonora and northern Sinaloa', *Technical Reports on the Meteorology and Climatology of Arid Regions*, no. 15 (University of Arizona, Institute of Atmospheric Physics).

[13] BENSON, LYMAN, and DARROW, R. A. (1954) *The Trees and Shrubs of the Southwestern Deserts* (Tucson: University of Arizona Press).

[14] ASCHMANN, HOMER (1959) 'The Central Desert of Baja California: demography and ecology', *Ibero-Americana*, 42 (Berkeley: University of California Press).

20 The Deserts of Dzungaria and the Tarim Basin

E. M. MURZAYEV

From *Nature of Sinkiang and Formation of the Deserts of Central Asia*, Joint Publications Research Service No. 40,299 (Washington, D.C., 1967) 327–30, 333, 337–9. Translated from *Priroda Sin'tszyana i Formirovaniye Pustyn' Tsentral'noy Azii*, published by 'Nauka' under auspices of the Institute of Geography of the U.S.S.R. Academy of Sciences (Moscow, 1966).

DZUNGARIA

THE saxaul desert is the most widely distributed type in Dzungaria; it is found not only on the sandy substratum, but also occupies regions of clayey or stony hammadas, both in the east and in the west, extending from Lake Ebi-Nuur to the upper part of the Urungu–Chernyy Irtysh interfluve. Here two types of saxaul grow: white saxaul (*Haloxylon persicum*) and Zaysan saxaul (*H. ammodendron*). The former prefers to grow on sand, as it does in Middle Asia. Dzungaria is the extreme eastern limit of its distribution. We observed white saxaul in the Kobbe ergs at the boundary with the Suluku landmark, i.e. already in the extreme north, near the Urungu valley. It is also common in the P'ich'ang ergs (eastern part of Dzungaria), being supplemented by selin [a grass] (*Aristida pennata*), and in the depressions between the ridges, with tamarisk, reaumuria and winterfat.

The Zaysan saxaul is very widely distributed on the rubbly pebble surfaces with the skeletal grey-brown soils of the desert. As is well known, this is the Gobi variety, which forms the landscapes of the stony deserts of Mongolia, where there is no white saxaul. On the sands fringing the northern, Tarim basin part of the Takla-Makan, and in individual sandy depressions on the left bank of the Tarim, specimens of this species of saxaul may be encountered, but here, as it appears, lies the extreme limit of its habitat. On the Kunlun plain, there are none of these plants (Fig. 20.1).

Together with the Zaysan saxaul, on these same stony soils rare clumps of such exceptionally hardy xerophytes as ephedra (*Ephedra przewalskii*) and iljinia (*Iljinia regelii*) may be found growing, associated with the gypsum-bearing soils.

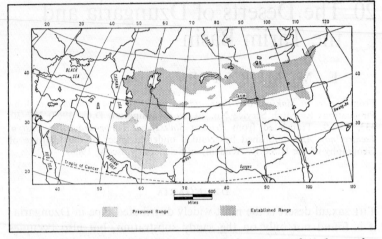

Fig. 20.1 Habitat of central Asiatic–Iranian–Turanian group of woody saxauls – Haloxylon persicum, H. aphyllum *and* H. ammodendron *(according to Ye. M. Lavrenko [1])*

On the sands of Dzungaria, besides these two species of saxaul such Turanian plants as selin, santolina feather-grass (*Artemisia santolina*), and also typical Gobi dwellers such as Mongolian calligonum (*Calligonum mongolicum*) and hedysarum (*Hedysarum mongolicum*) are found.

N. M. Przheval'skiy [2] in his traverse of Dzungaria also noted broomrape (*Orobanche salsa*), cynomorium (*Cynomorium songoricum*), which is well known to the Mongols under the name of 'goyo', and in the depressions between the ridges he noted rhubarb (*Rheum nanum*), and tulips, among the ephemeroids.

At the boundary between the accumulative-alluvial plains and the sandy deserts, in the strip where all the rainwater or the residual river waters are discharged and lie stagnant, a considerable salinisation of the soil mantle is apparent. In such places Zaysan saxaul dominates, growing together with *Reaumuria songorica*, tamarisk and winterfat (*Eurotia ceratoides*).

Extensive areas of accumulative plains in Dzungaria are occupied by growths of low hardy reaumuria shrubs, which create the yellowish-blue background of the desert. They do not avoid some salinisation of the dusty or rubbly soils. In small sinkholes, reaumuria forms communities with annual thistles.

These unique deserts are found in the northern part of the Dzungarian plains, where on the flat Tertiary kyrs, with clearly expressed microrelief and mesorelief of the residual formations [i.e. buttes and mesas] and the basins of considerable area the soils are comparatively slightly salinised and are formed on a rubbly-pebbly, sandy or clayey substratum. Here the desert saxaul–biyurgun association dominates, which has common features with the deserts of Kazakhstan and the south-western part of Mongolia. Besides the saxaul and Kazakhstan thistle or biyurgun (*Anabasis salsa*), other plants present here include winterfat, Shishkin's sagebrush (*Artemisia schischkinii*), *A. gracilescens*, *A. terrae albae* var. *massagetovii*, keurek (*Salsola rigida*), and, on the sandy places, sand sagebrush (*Artemisia arenaria*). A large umbellifer, ferula (*Ferula* sp.) is also found. In depressions tas-biyurgun (*Nanophyton erinaceum*) grows, as well as *Ephedra przewalskii* and a small quantity of ephemers.

The deserts of Dzungaria which occupy the lowest position are part of the Ebi-Nuur basin; the highest are found in the eastern part of the Dzungarian basin, where uplifted plains between the river Urungu and the Kuch'eng region are located. Both in the west and in the east the deserts are quite clearly expressed, but their botanico-geographical characteristics do not repeat each other.

In the Ebi-Nuur basin, sands are developed to a considerable degree, as well as solonchaks and piedmont boulder trains. Here small saxauls with *Ephedra przewalskii* dominate, together with iljinia, tas-biyurgun and Borotala sagebrush (*Artemisia borotalensis*). On the low part of the boulder trains, short-leaved cockspur (*Anabasis brevifolia*) predominates, and this is also common in the Mongolian Gobi.

The Tertiary plateaux with their incised valleys and basins extend into the east of Dzungaria where they are replaced by inclined plains with proluvial rubbly mantles, also dissected by gullies and valleys, in places with saline bottoms. In the depressions tamarisks and an endemic Mongolian brachanthemum (*Brachanthemum mongolicum*) are common, while on the hammadas are found *Ephedra przewalskii*, and bean-caper (*Zygophyllum xanthoxylon*), which is also widely distributed in the Zaaltayskaya (Trans-Altaic) Gobi, and saxaul on the more unconsolidated substratum. As we have already noted, the habitat of white saxaul, which forms considerable communities in the Kuch'eng sands, also extends to eastern Dzungaria, which, of course, is very curious since this species is most strongly associated with the Turanian lowlands and is already not encountered in the east of Central Asia. In any case, it is not observed in the neighbouring Peishan. On the dense clayey or stony mantles, in places somewhat mixed with sand, one may observe thistle (*Nitraria sphaerocarpa*), peatree (*Caragana leucophloae*), kurchavka (*Atraphaxis compacta*), Russian thistle, seepweed and summer cypress [2].

TARIM BASIN

The soil mantle and vegetation cover of the Tarim basin have many specific features. The extreme desert conditions, we may say the extra aridity, and the degree to which the basin is cut off by the highest mountains from the surrounding regions, could not help but facilitate an impoverishment of the flora: the number of species in southern Sinkiang is only a few hundred altogether. Here some species proper to Middle Asia or characteristic of it are also observed, but simultaneously such Central Asiatic endemics as Przheval'skiy's bean-caper, Kashgar reaumuria, Roborovskiy's limonium, and tansy (*Tanacetum brachanthemoides* and *T. kaschgaricum*) are represented.

A considerable area of the Tarim basin is occupied by sands not overgrown with any vegetation, the material of which is constantly blown about by the winds. In practice, there are no soils here, since soil-forming processes are absent.

The vegetation of the plains in the Tarim basin is exceptionally poor in species, among which, in the opinion of M. G. Popov [3], there are no new endemic forms and fresh species-forming impulses are lacking. The composition of the flora is Central Asiatic, although Turanian elements are also observed in it because of the proximity of Middle Asia. It is interesting to note that in the deserts on the mountain slopes surrounding the basin this author sees clear relics of the Pre-Miocene (Lower Tertiary) desert flora. Such forms, in particular as ephedra, anabasis, Kashgar reaumuria, false tamarisk (*Myricaria* sp.), nitraria, bean-caper, etc., are included. M. G. Popov notes the total absence of psammophytes in the sands of the Takla-Makan (by which he distinguishes them from the sands of the Turanian lowlands and Dzungaria); here we find no woody thistle such as cherkez (*Salsola richteri*), calligonum, or even selin. Only on the sands of the piedmont plain, where the climatic conditions and the water regime are somewhat more favourable, is the Gobi kumarchik and tickseed (*Corispermum* sp.).

Along the northern edges of the Takla-Makan, A. A. Yunatov [4] notes the Zaysan saxaul, sand sagebrush, camel thorn and karelinia, located in the depressions between the sand ridges with signs of salinisation. On the southern edge of this sand massif, where the ground water lies close to the surface, silt hills overgrown with tamarisk and heterophyllous poplar are prominent. At one time these poplars [cottonwoods, aspens] occupied considerably larger areas, but in our times their diminished habitats have become constricted as a result of various causes. Here we will note only one reason – the activity of man in the oases located in the higher part of the Tarim basin, under the Kunlun mountains.

The central part of the sands of the Takla-Makan is practically deprived of any vegetation, if we ignore the old riverbeds intersecting it, where it is possible to encounter fragments of tugays, composed chiefly of tamarisk and turanga. In the depressions between the ridges sometimes the same species are noted, to which certain thistles are added from time to time. On the sandy and rubbly proluvial deserts or hammadas (with gypsum-bearing soils over considerable areas) surrounding the alluvial plains of the Tarim basin, the composition of the vegetation is extremely monotonous; it is represented by thinly scattered shrubs, which can endure the excessive aridity. Such plants include *Ephedra przewalskii*, selitryanka (*Nitraria sphaerocarpa*), iljinia (*Iljinia regelii*) and sometimes Roborovskiy's calligonum.

M. G. Popov considers the following plants as being most characteristic for the rubbly-sandy deserts of southern Sinkiang: the species of ephedra, calligonum, iljinia and selitryanka just mentioned, and also the Siberian selitryanka, leafless yezhovik or itsegek [i.e. *Anabasis aphylla*], sympegma (*Sympegma regelii*), and in the east, at the foot of the Tien-Shan, Zaysan saxaul, *Kochia divaricata* and *Halogeton glomeratus* are added to them.

'On the plains of southern Sinkiang', writes A. A. Yunatov, 'cases are noted when over the extent of dozens of kilometres of travel a total lack of higher perennial plants on the rubbly surfaces (the rock trains of the Toksun basin, the Kunlun between Kargalyk and Khotan, and the Gashun Gobi) can be found. It is possible that in exceptionally rainy years annual, long-growing thistles (chiefly *Halogeton glomeratus*) may develop; as for the lower plants living in the soil, as yet we can say nothing' [4].

REFERENCES

[1] LAVRENKO, YE. M. (1962) 'Basic features of botanical geography of the deserts of Eurasia and North Africa', *Izd-vo AN SSSR, Komarovskiye chteniya* (Publications of the U.S.S.R. Academy of Sciences, Komarov Readings) (Moscow–Leningrad) xv.

[2] PRZHEVAL'SKIY, N. M. (1948) Moscow, Geografgiz. Originally in *Iz Zaysana cherez Khami v Tibet i na verkhov'ya Zheltoy reki* (From Zaysan through Hami to Tibet and to the Upper Reaches of the Yellow River) (St Petersburg, 1883).

[3] POPOV, M. G. (1958) 'Between Mongolia and Iran', *Izbrannyye Sochineniya* (Selected Works) (Ashkhabad, Izd-vo AN Turkmenistan SSR).

[4] YUNATOV, A. A. (1960) 'On several ecological and geographical regularities of the vegetation cover of the Sinkiang-Uighur Autonomous Region', in *Prirodnyye usloviya Sin'tszyana* (Natural Conditions of Sinkiang) (Moscow, Izd-vo AN SSSR).

21 Vegetational Types of Polar Lands

N. POLUNIN

From *Introduction to Plant Geography* (Longmans, London, 1960) pp. 382–95.

UNDER the prevailing cool conditions, water is very widely sufficient in the Arctic for such limited growth as the climate, etc., allows, and the main vegetational differences in any particular belt are rather in accordance with the actual habitats. Thus local edaphic or physiographic variations can ring the most immediate and fundamental changes in the local plant life. On the other hand a progressive and almost regular over-all depauperation of the vegetation is to be observed as we go farther and farther north; and as this tends to be rather closely comparable in the various sectors, it is deemed expedient to separate each sector (and consequently the Arctic as a whole) roughly into three main belts. These are the *low-Arctic*, in which the vegetation is continuous over most areas, the *middle-Arctic*, in which it is still sufficient to be widely evident from a distance, covering most lowlands, and the *high-Arctic*, in which closed vegetation is limited to the most favourable habitats and is rarely at all extensive. The following outline account of the main vegetational types of the Arctic will accordingly have, under each major heading, some consideration of the expression of this type in each of these three belts, ranging from south to north. Examples of low-arctic lands are the southern portions of almost all sectors, of middle-arctic lands Jan Mayen Island and the vicinity of Point Barrow, Alaska, and of high-arctic lands the whole of the Spitsbergen Archipelago, and the Canadian Eastern Arctic north of Lancaster Sound.

ARCTIC TUNDRAS

The term 'tundra', meaning essentially a treeless plain, has been used in so many and often such vague senses that it seems desirable, if we are to retain it at all, to limit its use so that it will have a more precise scientific connotation. In the present work the tundra proper is understood as the usually 'grassy' formation lying beyond (or in some extra-arctic places forming patches within) the limit of arborescent growth – except where shrubs or undershrubs predominate (in scrub and heathlands), or vegetation covers less than half of the area (in 'fell-fields' and 'barrens'). Generally comparable types occur in antarctic regions where, however, suitable land areas are relatively small. Alpine tundra bears a similar relationship to the timber-line on mountains. Instead of true grasses which, however, are rarely absent, grass-like plants such as sedges (*Carex* spp.), cotton-grasses (*Eriophorum* spp.), rushes (*Juncus* spp.) and wood-rushes *Luzula* spp.) commonly afford most of the 'grassiness' of the tundra, though various perennial forbs are usually associated, as are often a sprinkling of dwarf woody plants.

Even in this restricted sense the tundra developed in almost any arctic region is usually very variable, different areas supporting widely different types. The variation takes place particularly with differences in exposure and in water and other soil conditions, and, at all events in low- and middle-arctic regions, affords faciations far too numerous even to mention here. We may, however, distinguish and outline, besides a general central type, the tundras of damper depressions on the one hand and of drier exposed areas on the other.

The general run of tundra which covers a large proportion of the lowland plains and some less extensive upland areas of most low-arctic regions is commonly a rather thin 'grassy' sward dominated by mesophytic sedges such as the rigid sedge (*Carex bigelowii* agg.) and grasses such as the arctic meadow-grass (*Poa arctica* s.l.), with various associated forbs and undershrubs including dwarf willows. The whole forms a continuous if often poor sward commonly 15–35 cm. (approximately 6–14 in.) high in which a mixture of various bryophytes and lichens usually forms a rather poorly marked second layer a very few centimetres high.

Commonly the low-arctic tundra is a mosaic made up of faciations having each some lesser number of the total association dominants, and including consociations having only one of these. The areas of the component communities are often small and the variation from spot to spot in the tundra is accordingly usually considerable. In addition there are often local societies dominated by species other than the association dominants. The (sometimes unaccountable) mixing and even intergradation of all these communities is often intricate and may be suggestive of their relative youth, many having apparently failed to come to a state approaching equilibrium with the environment since emergence from glaciation or other extreme disturbance. Actually, it may be questioned whether, in many areas, even relative equilibrium can be attained in the face of the persistent frost-activity, and it has been claimed that the whole system constitutes an 'open' one in which the main tendency is repeated readjustment to almost perpetual disturbance.

With the generally poor drainage resulting from the soil being permanently frozen to not far beneath the surface, damper depressions or marshy open tracts tend to be plentiful although often of quite limited extent; indeed they are rarely absent except in regions of porous substrata and low water-table. In the low-Arctic they are commonly rather luxuriantly vegetated, the sward often being taller than it is in drier areas. They are usually dominated by cottongrasses and relatively hygrophytic sedges such as marshland ecads of the water sedge (*Carex aquatilis* agg.), and by grasses such as the arctagrostis (*Arctagrostis latifolia* s.l.), with a few hygrophilous willows or other ground-shrubs and many hygrophilous or ubiquitous forbs. Typical among these last are viviparous knotweed (*Polygonum viviparum*) and the bright-flowered yellow marsh saxifrage (*Saxifraga hirculus* agg.). The fairly luxuriant cryptogamic layer is largely composed of mosses, and helps to consolidate the whole. Often these marshy areas are beset with small hummocks commonly about 25 cm. high, and introducing drier conditions on their tops, which may then support healthy plants and lichens. Such hummocky tracts are known as 'hillock tundra'. In other instances tundras, especially of the damper types, are liable to be much interrupted by various of the geodynamic influences prevalent in cold regions – such as, particularly, solifluction and 'patterned soil' (polygon) formation of various kinds.

The drier tundras of raised areas or well-drained surface material in low-arctic regions tend to be much poorer and thinner than the damper types. Typically they are composed of an extremely various array of more or less xerophilous sedges (such as the rock sedge, *Carex rupestris*), willows (particularly the arctic willow, *Salix arctica* s.l.), grasses (such as alpine holy-grass, *Hierochloe alpina*), northern wood-rush (*Luzula confusa* agg.), and various forbs (such as the same viviparous knotweed), in addition to mountain and arctic avens (*Dryas* spp.), which are somewhat woody, and which may dominate considerable areas. But although scattered heathy plants occur in them, these areas are scarcely heaths, any more than are the lichen-rich ones dominated by xerophilous sedges that characterise dry and exposed situations. Moreover their vegetation is usually rather poor, often barely covering the ground in spite of a plentiful admixture of lichens and sometimes also of bryophytes. Plate 14 shows such an area in which boulders project through the thin and somewhat heathy, lichen-rich vegetation. Especially on limestone or porous sandy substrata is growth often poor and the vegetation relatively sparse, although the component flora particularly in calcareous areas may be very various.

It would accordingly seem that the major variations in the precise type of tundra take place chiefly, but by no means solely, with local water conditions working through exposure or edaphic factors, while very locally the effect of frost action may be paramount. Thus differences in substratum, as between limestone and acid-weathering rock, can introduce vegetational differences due to particular plants' preferences quite apart from water-relations, while, as an example of the entry of another factor, heavy pasturing can lead to increased grassiness as in temperate regions. In addition, polygon-formation and solifluction may cause persistent disturbance. It may be noted that, whereas the dominants are usually at least specifically distinct in different types of low-arctic tundra, some of the less exacting, more tolerant associates may be present in a wide range of habitat types. This again is comparable with the situation in cool-temperate regions and, it often seems, obtains still more forcibly to the north. Thus in the Far North some of the hardier plants, such as viviparous knotweed and some of the saxifrages, grow in an extraordinarily wide variety of habitats, ranging from wet to dry, exposed to sheltered, and open-soil to vegetationally 'closed'.

The middle-arctic belt is characterised by tundras of a generally poorer type, both in the matter of flora and luxuriance of development, than the low-arctic ones. Thus some of the plants which were important in low-arctic tundras are absent, though all of the dominants, etc., mentioned above for low-arctic tundras can, and frequently do, occupy a similar position also in middle-arctic regions. Moreover the range of types is much the same, damp, mesophytic and drier ones being distinguishable.

In high-arctic regions still further depauperation is general, and indeed only limited and relatively few areas are sufficiently vegetated to be designated as tundra. These areas are chiefly marshy ones and may be still dominated by sedges, cotton-grasses and grasses – often of the same species as in the south, and including similar associated forbs, though woody plants apart from prostrate willows are usually absent. Mosses commonly consolidate the whole, and in some places appear to dominate. While the main dominants in such areas are commonly sedges and grasses growing on the sides or tops of the hummocks, the microhabitat effect is extreme, the microhabitats ranging from depressions occupied by dark boglets or puddles of 'free' water to dry hillock tops occupied by lichens or, in favourably sheltered situations, xeromorphic ground-shrubs.

Tracts of 'grassy' mesophytic tundra of any substantial extent are not common in the high-Arctic. Drier types of tundra in these farthest north lands tend to be dominated largely by lichens and to be much interrupted by rocks or bare patches – especially in exposed situations.

So far as regular ecological successions are concerned, these are especially problematical in the Arctic. It has, however, been suggested that the marshy and dry tundras may be subclimax and the mesophytic ones climax or preclimax, the scrub and heathlands, which are developed in the most favourable situations (see below), being either postclimax or, perhaps, indicative of a more general climax to be expected ultimately in sufficiently favourable situations, though at present a mixed 'polyclimax' is commonly found. The significance of different 'stages' in the hypothetical successions may, however, vary from place to place. Thus, in the Far North, heathy plants are apt to be so restricted to the most favourable situations as to suggest that without major climatic change they could not become widely dominant in the manner already obtaining in some places in the low-Arctic. Moreover, frost and other disturbance is so widespread, *inter alia* impeding or even preventing the maturation of soils, that it seems as though many areas undergo a kind of perpetual readjustment rather than exhibit the tendency to equilbrium which is implicit in a real climax.

ARCTIC SCRUB AND HEATHLANDS

A shaggy scrub of willows and/or birches is commonly developed on the most favourable slopes, in damp depressions, and especially along watercourses and the margins of lakes in low-arctic regions.

It is commonly around 60 cm. (about 2 ft) high, as in the example shown in Plate 15, but tends to become lower and more restricted northwards until, about the centre of the middle-arctic belt, it becomes usually very limited in extent and stature. However, in the most favourable situations in the extreme south the Willows may be luxuriant and even exceed the height of a man, and especially in south-western Greenland the scrub is quite extensively developed, in some places including arborescent birches. These Greenland birch 'forests' are of very limited extent, with the trees scattered and scraggy though sometimes nearly 6 m. in height and 25 cm. in stem diameter. Their areas have been termed subarctic but seem too limited to separate on an over-all, world basis; they are also too fickle, the development of an arborescent habit being evidently dependent on local shelter, etc. Apart from these larger birches, the main dominants in different regions are most often the dwarf birch (*Betula nana* agg.) or scrub birch (*B. glandulosa* agg.), or such shrubby willows as the glaucous willow (*Salix glauca* s.l.), the broad-leafed willow (*S. cordifolia* s.l.), the feltleaf willow (*S. alaxensis* agg.), or Richardson's willow (*S. richardsonii* agg.). Often two or more of these scrubs will dominate a mixed association. In some places bushes of green alder (*Alnus crispa* agg.) are present and may be locally dominant.

Such scrub at its best is so thickly tangled and produces so much litter that few associated plants occur, apart from tall grasses such as bluejoint (*Calamagrostis canadensis* agg.) and occasional straggling forbs. But where the dominants are less luxuriant, an extensive flora is often found, including a considerable variety of herbs and mosses, or, in dry situations, subdominant heathy plants such as crowberry (*Empetrum nigrum* s.l.). Also characteristic of dry scrub are patches of tall cladonias, stereocaulons and other lichens, with or without polytricha or other coarse mosses. To the north such scrub thins out gradually, its most northerly expression about the northern limit of the middle-arctic belt being usually in the form of single or scarcely confluent bushes that rarely exceed 50 cm. in height and are usually much lower, though often quite wide.

Heathlands are more widespread and various in the Arctic than is scrub, though still commonly occupying only a very small proportion of the total area. They are usually characterised by being dominated by members of the heath family (Ericaceae) or by heath-like plants such as, particularly, crowberry. Sometimes, however, broad-leafed plants such as avens (*Dryas*), or sedges such as the nard sedge (*Carex nardina* s.l.) or Bellard's kobresia (*Kobresia myosuroides*), may dominate dry and usually exposed, lichen-rich areas that are often classed as heathlands rather than among the drier tundras with which they seem more properly to belong (see above). Leaving aside such cases it may be said that heathlands in the Arctic tend to be confined to the more favourable, sheltered situations that are snow-covered in winter – provided they are not too moist in summer. In many regions they characterise coarse-grained rather than clayey soils, as pointed out by Professor Thorvald Sørensen (*in litt.*).

In the low-arctic belt the heathlands are usually covered by a continuous thick sward of mixed woody and herbaceous plants, the main dominants being typically 8–15 cm. high. These commonly include crowberry, arctic blueberry (*Vaccinium uliginosum* subsp. *alpinum*), mountain cranberry (*V. vitis-idaea* agg.), arctic bell-heather (*Cassiope tetragona*), narrow-leafed labrador-tea (*Ledum palustre* agg.), dwarf birch and various diminutive willows. Often the dominants themselves are much mixed, and usually they are consolidated below by a layer of cryptogams in which mosses or lichens commonly subdominate according to whether the situation is relatively moist or dry, respectively. In the drier situations there may occur frequent gaps in the heath which are actually dominated by lichens – particularly by 'caribou-moss' cladonias that may form a sward 5 or more cm. high. In depressions and behind obstructions where snow drifts deeply in winter, a characteristic dark (except when flowering) heath dominated by arctic bell-heather usually develops, often with associated sedges and mosses at least where the soil is lastingly moist. Such an area is shown in Plate 16 and, apart from a zone of more mixed heath that may develop outside, usually constitutes the outermost of a zoned series of subclimaxes developed in late-snow areas.

In the middle-arctic belt, heathlands are usually somewhat lower in stature and more restricted in area than to the south, having the appearance of postclimaxes developed in the most favourable situations. Of the cited dominants mountain cranberry has usually disappeared, and although the taller ones may still exceed 20 cm. in height the sward is usually only 5–10 cm. high. Whereas it may still be fairly dense, more often the 'heath' is of scattered ground-shrubs with intervening thin patches of cetrarias, alectorias and other lichens.

Selected Bibliography

TROPICAL RAIN FOREST

AUBRÉVILLE, A. (1938) 'La forêt coloniale: les forêts de l'Afrique occidentale française', *Ann. Acad. Sci. Colon., Paris*, IX 1–245.

BEARD, J. S. (1945–6) 'The Mora Forests of Trinidad, British West Indies', *J. Ecol.*, XXXIII 173–92.

DAVIS, T. A. W., and RICHARDS, P. W. (1933) 'The vegetation of Moraballi Creek, British Guiana, Pt. I', *J. Ecol.*, XXI 350–84.

—— and —— (1934) 'The vegetation of Moraballi Creek, British Guiana, Pt. II', *J. Ecol.*, XXII 106–55.

EGGELING, W. J. (1947) 'Observations on the ecology of the Budongo rain forest, Uganda', *J. Ecol.*, XXXIV 20–87.

JONES, E. W. (1955–6) 'Ecological studies on the rain forest of southern Nigeria: IV. The plateau forest of the Okomu Forest Reserve', *J. Ecol.*, XLIII 564–94; XLIV 83–117.

KENOYER, L. A. (1929) 'General and successional ecology of the lower tropical rain-forest at Barro Colorado Island, Panama', *Ecology*, X 201–22

MOONEY, J. W. C. (1961) 'Classification of the vegetation of the high forest zone of Ghana', *UNESCO, Humid Tropics Res., Proc. of the Abidjan Symp., 1959*, pp.85–6.

POORE, M. E. D. (1963) 'Problems in the classification of tropical rain forest', *J. Trop. Geog.*, XVII 12–19.

RICHARDS, P. W. (1939) 'Ecological studies on the rain forest of Southern Nigeria: I. Primary forest', *J. Ecol.*, XXVII 1–61.

—— (1952) *The Tropical Rain Forest: An Ecological Study* (Cambridge).

ROSS, R. (1954) 'Ecological studies on the rain forest of Southern Nigeria: III. Secondary succession in the Shasha Forest Reserve', *J. Ecol.*, XLII 259–82.

TROPICAL SEASONAL FOREST AND TROPICAL SAVANNA

AIRY-SHAW, H. K. (1947) 'The vegetation of Angola', *J. Ecol.*, XXXV 23–48.

ASPREY, G. F., and LOVELESS, A. R. (1958) 'The dry evergreen formations of Jamaica, II', *J. Ecol.*, XLVI 547–70.

BARTLETT, H. H. (1956) 'Fire, primitive agriculture and grazing in the tropics', in W. L. Thomas (ed.), *Man's Role in Changing the Face of the Earth*, pp. 692–720.

BEARD, J. S. (1946) 'The natural vegetation of Trinidad', *Oxf. For. Mems.*, 20 (152 pp.).

BUXTON, P. A. (1935) 'Seasonal changes in vegetation in the north of Nigeria', *J. Ecol.*, XXIII 134–9.

CLAYTON, W. D. (1958) 'Secondary vegetation and the transition to savanna near Ibadan, Nigeria', *J. Ecol.*, XLVI 217–38.

COLE, M. M. (1960) 'Cerrado, Caatinga and Pantanal: the distribution and origin of the savanna vegetation of Brazil', *Geog. J.*, CXXVI (2) 168–79.

—— (1963) 'Vegetation nomenclature and classification with particular reference to the savannas', *S. African Geog. J.*, XLV 3–14.

FERRI, M. G. (1961) 'Aspects of soil–water–plant relationships in connexion with some Brazilian types of vegetation', *UNESCO, Humid Tropics Res., Proc. of the Abidjan Symp., 1959*, pp. 103–9.

GRIEG-SMITH, P. (1952) 'Ecological observations on degraded and secondary forest in Trinidad, British West Indies. I', *J. Ecol.*, XL 283–315.

KEAY, R. W. J. (1949) 'An example of Sudan zone vegetation in Nigeria', *J. Ecol.*, XXXVII 335–64.

LOVELESS, A. R., and ASPREY, G. F. (1957) 'The dry evergreen formations of Jamaica. I', *J. Ecol.*, XLV 799–822.

MORISON, C. G. T., HOYLE, A. C., and HOPE-SIMPSON, J. P. (1948) 'Tropical soil–vegetation catenas and mosaics: A study in the south-western part of the Anglo-Egyptian Sudan', *J. Ecol.*, XXXVI 1–84.

MYERS, J. G. (1933) 'Notes on the vegetation of the Venezuelan Llanos', *J. Ecol.*, XXI 335–49.

—— (1936) 'Savannah and forest vegetation of the interior Guiana Plateau', *J. Ecol.*, XXIV 162–84.

RAWITSCHER, F. (1948) 'The water economy of the vegetation of the "Campos Cerrados" in southern Brazil', *J. Ecol.*, XXXVI 237–68.

RICHARDS, P. W. (1952) *The Tropical Rain Forest: An Ecological Study*, chap. 15, pp. 315–45.

TRAPNELL, C. G. (1959) 'Ecological results of woodland burning experiments in Northern Rhodesia', *J. Ecol.*, XLVII 129–68.

TROPICAL DESERT AND SEMI-DESERT

DAVIS, P. H. (1953) 'The vegetation of the deserts near Cairo', *J. Ecol.*, XLI 157–73.

GILLILAND, H. B. (1952) 'The vegetation of eastern British Somaliland', *J. Ecol.*, XL 91–124.

KASSAS, M. (1952–3) 'Habitat and plant communities in the Egyptian Desert', *J. Ecol.*, XL 342–51; XLI 248–56.

—— (1956) 'The Mist Oasis of Erkwit, Sudan', *J. Ecol.*, XLIV 180–94.

—— and IMAM, M. (1954) 'Habitat and plant communities in the Egyptian Desert: III. The wadi bed ecosystem', *J. Ecol.*, XLII 424–41.

—— and —— (1959) 'Habitat and plant communities in the Egyptian deserts: IV. The gravel desert', *J. Ecol.*, XLVII 289–310.

VESEY-FITZGERALD, D. F. (1955) 'Vegetation of the Red Sea coast south of Jedda, Saudi Arabia', *J. Ecol.*, XLIII 477–89.

—— (1957) 'Vegetation of the Red Sea coast north of Jedda, Saudi Arabia', *J. Ecol.*, XLV 547–62.

—— (1957) 'The vegetation of central and eastern Arabia', *J. Ecol.*, XLV 779–98.

TROPICAL MONTANE VEGETATION

JACKSON, J. K. (1956) 'Vegetation of the Imatong Mountains, Sudan', *J. Ecol.*, XLIV 341–74.

RICHARDS, P. W. (1936) 'Ecological observations on the rain forest of Mount Dulit, Sarawak. I', *J. Ecol.*, XXIV 1–37.

—— (1936) 'Ecological observations on the rain forest of Mount Dulit, Sarawak. II', *J. Ecol.*, XXIV 340–60.

ROBBINS, R. G. (1961) 'The montane vegetation of New Guinea', *Tuatara*, VIII 3, 121–34.

SALT, G. (1954) 'A contribution to the ecology of upper Kilimanjaro', *J. Ecol.*, XLII 375–423.

SHREVE, F. (1914) 'A montane rain forest: a contribution to the physiological plant geography of Jamaica', *Carnegie Inst.*, 199 (pp. 110).

VAUGHAN, R. E., and WIEHE, P. O. (1941) 'Studies on the vegetation of Mauritius: III. The structure and development of the upland climax forest', *J. Ecol.*, XXIX 127–60.

MID-LATITUDE FORESTS

BRAUN, E. LUCY (1936) 'Forests of the Illinoian till plain of south-western Ohio', *Ecol. Mon.*, VI 89–149.

—— (1938) 'Deciduous forest climaxes', *Ecology*, XIX 515–22.

—— (1947) 'Development of the deciduous forests of eastern North America', *Ecol. Mon.*, XVII 211–19.

BROMLEY, S. W. (1935) 'The original forest types of southern New England', *Ecol. Mon.*, V 61–90.

BURGES, ALAN, and JOHNSTON, R. D. (1953) 'The structure of a New South Wales subtropical rain forest', *J. Ecol.*, XLI 72–83.

CHAPMAN, H. H. (1932) 'Is the longleaf type a climax?', *Ecology*, XIII 328–34.

CHAPMAN, V. J. (1958) 'The geographical status of New Zealand lowland forest vegetation', *New Zealand Geographer*, XIV 103–14.

COCKAYNE, L. (1926) 'Monograph on New Zealand beech forests', *N.Z. State Forest Bull.*, 4.

DAUBENMIRE, R. F. (1936) 'The "Big Woods" of Minnesota', *Ecol. Mon.*, VI 233–68.

DAWSON, J. W. (1962) 'The New Zealand lowland podocarp forest: Is it subtropical?' *Tuatara*, IX 98–116.

EGGLER, W. A. (1938) 'The maple-basswood forest type in Washburn County, Wisconsin', *Ecology*, XIX 243–63.

GULISASHVILI, V. Z. (1961) 'Sub-tropical forests of Transcaucasia', *UNESCO, Humid Tropics Res., Proc. of the Abidjan Symp., 1959*, pp. 69–74.

HALLIDAY, W. E. D. (1937) 'A forest classification for Canada', *Dept. of Research and Development, Forest Research Division*, Bulletin 89 (Ottawa).

HANSEN, H. P. (1938) 'Postglacial forest succession and climate in the Puget Sound region', *Ecology*, XIX 528–42.

KELLER, BORIS A. (1927) 'Distribution of vegetation on the plains of European Russia', *J. Ecol.*, XV 201–15.

KÜCHLER, A. W. (1946) 'The broadleaf deciduous forests of the Pacific Northwest', *A.A.A.G.*, XXXVI 122–47.

KWEI-LING CHU and COOPER, W. S. (1950) 'An ecological reconnais-sance in the native home of *Metasequoia glyptostroboides*', *Ecology*, XXXI 260–78.

LITTLE, E. I. (1939) 'The vegetation of Caddo County canyons, Oklahoma', *Ecology*, XX 1–10.

LUTZ, H. J. (1930) 'The vegetation of Heart's Content, a virgin forest in northwestern Pennsylvania', *Ecology*, XI 1–29.

MAYCOCK, P. F., and CURTIS, J. T. (1960) 'The phytosociology of boreal conifer-hardwood forests of the Great Lakes region', *Ecol. Mon.*, XXX 1–36.

MUNGER, T. T. (1940) 'The cycle from Douglas fir to hemlock', *Ecology*, XXI 451–9.

NICHOLS, G. E. (1935) 'The hemlock–white pine–northern hardwood region of eastern North America, *Ecology*, XVI 403–22.

PESSIN, L. J. (1933) 'Forest associations in the uplands of the lower Gulf Coastal Plain', *Ecology*, XIV 1–14.

ROBBINS, R. G. (1962) 'The podocarp–broadleaf forests of New Zealand', *Trans. Roy. Soc. N.Z.*, I 5.

RÜBEL, E. A. (1914) 'The forests of the western Caucasus', *J. Ecol.*, II 39–42.

STEARNS, F. W. (1949) 'Ninety years change in a northern hardwood forest in Wisconsin', *Ecology*, XXX 350–8.

—— (1950) 'The composition of a remnant of white pine forest in the Lake States', *Ecology*, XXXI 290–2.

YOUNG, V. A. (1933) 'Hardwood invasion in a comparatively old white pine afforested area', *Ecology*, XIV 61–9.

SCLEROPHYLLOUS SHRUBLANDS

ADAMSON, R. S. (1927) 'The plant communities of Table Mountain – account', *J. Ecol.*, XV 278–309.

—— (1931) 'The plant communities of Table Mountain – life-form' *J. Ecol.*, XIX 304–20.

—— (1935) 'The plant communities of Table Mountain – regeneration', *J. Ecol.*, XXIII 45–55.

BEARD, J. S. (1967) 'A study of patterns in some West Australian heath and mallee communities', *Aust. J. Bot.*, XV 131–9.

BURBIDGE, NANCY T. (1952) 'The significance of the mallee habit in Eucalyptus', *Proc. Roy. Soc. Queensland*, LXII 73–8.

COOPER, W. S. (1922) 'The broad-sclerophyll vegetation of California', *Carnegie Inst. Wash.*, 319.

GIMINGHAM, C. H., and WALTON, K. (1954) 'Environment and the structure of scrub communities on the limestone plateaux of northern Cyrenaica', *J. Ecol.*, XLII 505–20.

SCOTT-ELLIOT, G. F. (1889) 'Notes on the regional distribution of the Cape flora', *Trans. Edinburgh Botanical Society*, XVIII.

WOOD, J. G. (1929) 'Floristics and ecology of the mallee', *Trans. Roy. Soc. S. Aust.*, LIII 359–78.

MID-LATITUDE GRASSLANDS AND GRASSLAND–FOREST ECOTONES

ACOCKS, J. P. H. (1953) 'Veld types of South Africa', *Bot. Survey of South Africa*, Memoir 28.

ALBERTSON, F. W. (1937) 'Ecology of mixed prairie in west-central Kansas', *Ecol. Mon.*, VII 481–547.

BARKER, A. P. (1953) 'An ecological study of tussock grassland, Hunters Hills, South Canterbury', *N.Z.D.S.I.R.*, Bulletin 107.

BIRD, R. D. (1930) 'Biotic communities of the aspen parkland of central Canada', *Ecology*, XI 356–442.

BRUNER, W. E. (1931) 'The vegetation of Oklahoma', *Ecol. Mon.*, I 99–188.

BUELL, M. F., and CANTLON, J. E. (1951) 'A study of two forest stands in Minnesota with an interpretation of the prairie-forest margin', *Ecology*, XXXII 294–316.

COUPLAND, R. T. (1950) 'Ecology of mixed prairie in Canada', *Ecol. Mon.*, XX 271–316.

CUMBERLAND, K. B. (1962) 'Climatic change or cultural interference?' in *Land and Livelihood: Geographical Essays in Honour of George Jobberns* (Christchurch, N.Z.).

DYKSTERHUIS, E. V. (1948) 'The vegetation of the Western Cross Timbers', *Ecol. Mon.*, XVIII 325–76.

EWING, J. (1924) 'Plant successions of the bush-prairie in north-western Minnesota', *J. Ecol.*, XII 238–66.

GATES, F. C. (1926) 'Pines in the prairie', *Ecology*, VII 96–8.

GLEASON, H. A. (1909) 'Some unsolved problems of the Prairies', *Bull. Torr. Bot. Club*, XXXVI 265–71.

—— (1922) 'The vegetational history of the Middle West', *A.A.A.G.*, XII 39–85.

KELLER, BORIS A. (1927) 'Distribution of vegetation on the plains of European Russia', *J. Ecol.*, XV 215–18.

LARSON, F. (1940) 'The role of the bison in maintaining the short grass plains', *Ecology*, XXI 113–21.

—— and WHITMAN, W. (1942) 'A comparison of used and unused grassland mesas in the badlands of South Dakota', *Ecology*, XXIII 438–45.

MOSS, E. H. (1932) 'The vegetation of Alberta: IV. The poplar association and related vegetation of central Alberta', *J. Ecol.*, XX 380–415.

QUINNILD, C. L., and COSBY, H. E. (1958) 'Relicts of climax vegetation on two mesas in western North Dakota', *Ecology*, XXXIX 29–32.

ROSTLUND, E. (1957) 'The myth of a natural prairie belt in Alabama', *A.A.A.G.*, XLVII 392–411.

SCHMEIDER, O. (1927) 'The Pampa – a naturally or culturally-induced grassland?', *Univ. of Calif. Pubs. in Geog.*, II 255–70.

SHIMEK, B. (1911) 'The Prairies', *Bull. Lab. Nat. Hist. State Univ. Iowa*, VI 169–240.

—— (1925) 'Papers on the Prairie', *Univ. Iowa Studies in Nat. Hist.*, XI 5, 1–36.

SHULL, C. A. (1921) 'Some changes in the vegetation of western Kentucky', *Ecology*, II 120–4.

STEIGER, T. L. (1930) 'Structure of prairie vegetation', *Ecology*, XI 170–217.

STODDART, L. A. (1941) 'The Palouse grassland association in northern Utah', *Ecology*, XXII 158–63.

THOMSON, J. W. (1940) 'Relict prairie areas in central Wisconsin', *Ecol. Mon.*, X 685–717.

TRANSEAU, E. N. (1935) 'The Prairie Peninsula', *Ecology*, XVI 423–37.

WEAVER, J. E. (1943) 'Replacement of true prairie by mixed prairie in eastern Nebraska and Kansas', *Ecology*, XXIV 421–34.

—— and FITZPATRICK, J. (1934) 'The Prairie', *Ecol. Mon.*, IV 109–298.

MID-LATITUDE DESERT AND SEMI-DESERT

HANSON, H. C. (1924) 'A study of the vegetation of northeastern Arizona', *Univ. Neb. Studies*, XXIV 85–94.

HASTINGS, J. R., and TURNER, R. M. (1965) *The Changing Mile* (Tucson, Ariz.).

KELLER, BORIS A. (1927) 'Distribution of vegetation on the plains of European Russia', *J. Ecol.*, XV 218–26.

MacDougal, D. T. (1908) 'Botanical features of North American deserts', *Carnegie Inst. Wash.*, 99.

Murzayev, E. M. (1967) 'Nature of Sinkiang and formation of the deserts in central Asia', *Joint Publications Research Service*, no. 40, 299 (U.S. Dept. of Commerce, Washington, D.C.).

Parish, S. B. (1930) 'Vegetation of the Mohave and Colorado Deserts of southern California', *Ecology*, XI 481–99.

Shreve, F. (1915) 'The vegetation of a desert mountain range as conditioned by climatic factors', *Carnegie Inst. Wash.*, 217.

—— (1925) 'Ecological aspects of the deserts of California', *Ecology*, VI 93–103.

—— (1936) 'The plant life of the Sonoran Desert', *Sci. Monthly*, XLII 195–213.

TUNDRA AND TUNDRA–FOREST ECOTONES

Drew, J. V., and Shanks, R. E. (1965) 'Landscape relationships of soils and vegetation in the forest–tundra ecotone, Upper Firth Valley, Alaska–Canada', *Ecol. Mon.*, XXXV 285–306.

Falk, P. (1940) 'Further observations on the ecology of central Iceland', *J. Ecol.*, XXVIII 1–41.

Griggs, R. F. (1934) 'The edge of the forest in Alaska and the reasons for its position', *Ecology*, XV 80–96.

Keller, Boris A. (1927) 'Vegetation of the plains of European Russia', *J. Ecol.*, XV 191–201.

Leach, W., and Polunin, N. (1932) 'The vegetation of Finnmark', *J. Ecol.*, XX 416–30.

Marr, J. W. (1948) 'Ecology of the forest–tundra ecotone on the coast of Hudson Bay', *Ecol. Mon.*, XVIII 117–44.

Polunin, N. (1945–6) 'Plant life in Kongsfjord, West Spitzbergen', *J. Ecol.*, XXXIII 82–108.

—— (1951) 'The real Arctic: suggestions for its delimitation, subdivision and characterisation', *J. Ecol.*, XXXIX 308–15.

Russell, R. S., and Wellington, P. S. (1940) 'Physiological and ecological studies on an Arctic vegetation: I. The vegetation of Jan Mayen Island', *J. Ecol.*, XXVIII 153–79.

Tikhomirov, B. A. (1962) 'The treelessness of the tundra', *Polar Record*, XI 24–30.

REGIONAL ACCOUNTS

ADAMSON, R. S. (1938) *The Vegetation of South Africa.*

BEADLE, N. C. W. (1948) *The Vegetation and Pastures of Western New South Wales* (Sydney).

BEARD, J. S. (1944) 'The natural vegetation of the island of Tobago, British West Indies', *Ecol. Mon.*, XIV 135–64.

—— (1946) 'The natural vegetation of Trinidad', *Oxf. For. Mems.*, 20.

—— (1955) 'The classification of tropical American vegetation types', *Ecology*, XXXVI 89–100.

COCKAYNE, L. (1928) *Vegetation of New Zealand*, 2nd ed. (Leipzig).

DAWSON, J. W. (1963) 'New Caledonia and New Zealand – a botanical comparison', *Tuatara*, XI 178–93.

EDWARDS, D. C. (1940) 'A vegetation map of Kenya', *J. Ecol.*, XXVIII 377–85.

HERBERT, D. A. (1935) 'The climatic sifting of Australian vegetation', *Rep. Melb. Meeting, Aust. and N.Z. Ass. Adv. Sci.*, pp. 349–70.

KELLER, BORIS A. (1927) 'Distribution of vegetation on the plains of European Russia', *J. Ecol.*, XV 189–233.

LEOPOLD, A. S. (1950) 'Vegetation zones of Mexico', *Ecology*, XXXI 507–18.

SEIFRIZ, W. (1931) 'Sketches of the vegetation of some southern provinces of Soviet Russia, II', *J. Ecol.*, XIX 360–82.

—— (1932) 'Sketches of the vegetation of some southern provinces of Soviet Russia, III–V', *J. Ecol.*, XX 53–88.

—— (1935) 'Sketches of the vegetation of some southern provinces of Soviet Russia, VI–VII', *J. Ecol.*, XXIII 140–60.

SHREVE, F. (1915) 'The vegetation of a desert mountain range as conditioned by climatic factors', *Carnegie Inst.*, 217 (pp. 112).

SPECHT, R. L., and MOUNTFORD, C. P. (1958) 'Botany and plant ecology', in *Records of the American–Australian Scientific Expedition to Arnhem Land*, III (Melbourne).

STAMP, L. D. (1925) *The Vegetation of Burma* (Rangoon).

TUTIN, T. G. (1953) 'The vegetation of the Azores', *J. Ecol.*, XLI 53–61.

VAUGHAN, R. E., and WIEHE, P. O. (1937) 'Studies on the vegetation of Mauritius: I. A preliminary survey of the plant communities', *J. Ecol.*, XXV 289–343.

WEAVER, J. E., and CLEMENTS, F. E. (1938) *Plant Ecology* (New York) chap. XVIII, pp. 478–538.

WEBB, L. J. (1959) 'A physiognomic classification of Australian Rain Forests', *J. Ecol.*, XLVII 551–70.

ZOHARY, M. (1947) 'A vegetation map of western Palestine', *J. Ecol.*, XXXIV 1–19.